PETER LANG
PROMPT

PETER LANG
New York • Bern • Berlin
Brussels • Vienna • Oxford • Warsaw

Anatoly Nazirov

The Socio-Cultural and Philosophical Origins of Science

Translated from the Russian by
Ivan Zhavoronkov

PETER LANG
New York • Bern • Berlin
Brussels • Vienna • Oxford • Warsaw

Library of Congress Cataloging-in-Publication Data

Names: Nazirov, Anatoly, author. | Zhavoronkov, Ivan, translator.
Title: The socio-cultural and philosophical origins of science /
Anatoly Nazirov; translated from the Russian by Ivan Zhavoronkov.
Other titles: Sotsiokul'turnye i filosofskie istoki nauki. English
Description: New York: Peter Lang, 2020.
Includes bibliographical references and index.
Identifiers: LCCN 2019024949 | ISBN 978-1-4331-7228-1 (hardback: alk. paper)
ISBN 978-1-4331-7251-9 (ebook pdf)
ISBN 978-1-4331-7252-6 (epub) | ISBN 978-1-4331-7253-3 (mobi)
Subjects: LCSH: Science—Philosophy.
Classification: LCC Q175 .N36413 | DDC 501—dc23
LC record available at https://lccn.loc.gov/2019024949
DOI 10.3726/b15964

Bibliographic information published by **Die Deutsche Nationalbibliothek**.
Die Deutsche Nationalbibliothek lists this publication in the "Deutsche Nationalbibliografie"; detailed bibliographic data are available on the Internet at http://dnb.d-nb.de/.

The Russian original was published by
Saint Petersburg State University of Service and Economics Press in 2009

Permission to publish the present translation of the Russian original was obtained from
Saint Petersburg State University of Economics Press in 2019

This publication has been peer reviewed and meets
the highest quality standards for content and production.

The paper in this book meets the guidelines for permanence and durability
of the Committee on Production Guidelines for Book Longevity
of the Council of Library Resources.

29 Broadway, 18th floor, New York, NY 10006
www.peterlang.com

Printed in the United States of America

Contents

Acknowledgement vii
Introduction ix

1. The Psychosomatic Foundations of Spiritual Culture 1
2. The Attempt at a Pre-theoretical Synthesis of Knowledge in Theosophy 37
3. The Problem of Prerequisites for a Scientific Theory in the History of Philosophy 53

Conclusion 71
Bibliography 73
Name Index 79
Subject Index 83

Acknowledgement

Special thanks to the acquisition editor Meagan Simpson, the editorial assistant Liam McLean, the copyeditor Kondappan Sathiaseelan, and the production editors Jackie Pavlovic and Sasireka Sakthi for helping to make this book possible.

Introduction

This study discusses the formation of spiritual culture and reveals the prerequisites for the developments of philosophy, science, religion, and art from a common root—animistic thinking. Philosophy emerges as reflexive thinking which transforms the animistic into the ideal and forms the basis for the emergence and development of science. Philosophy as an element of spiritual culture—a culture which is involved in the process of the psychosomatic functioning of the human being—allows one to reflect on its own foundations and to understand the nature of the ideal, which is formed by logical instruments into what unfolds as philosophico-scientific rationality.

Chapter One

The Psychosomatic Foundations of Spiritual Culture

The formulation and discussion of the problem of the inter-relationship between spiritual culture and psychosomatics has become possible due to the expansion of the content and scope of the methodology of science that has occurred over the last decades. Such a change has been caused by the inclusion in the arsenal of methodological research of the question of the socio-cultural determination of the genesis of scientific knowledge—analyses of the relations between the rational and the irrational, the discursive and the intuitive, the naturo-scientific and the socio-human, and the cognitive and the valuative. This has led to the development and rethinking of the very concept of rationality, including the classification of its types, which has entailed the need to supplement rationality with the historical method of research.[1]

In the methodology of science, there are two opposite approaches to the question of the relationship between the rational and the irrational. The first—close to the traditional point of view—is that in the process of cognition the non-rational is constantly rationalized. This view is shared by philosophers who take the position of critical rationalism or stand close to it (e.g., J. Watkins). The second—non-traditional—approach recognizes the leading role of the irrational over the rational. In post-positivist philosophy, such a position is shared by supporters of historical schools (T. Kuhn, P. Feyerabend).

A third approach consists in the evaluation of the rational and the non-rational, not from the perspective of their individual specifics, but "primarily from the point of view of the nature of knowledge."[2] In this case, the discussion of the problem of the methodological analysis of knowledge is transferred to the context of the latter's socio-cultural functioning, one of the essential aspects of which is imagination. "Any underestimation of the socio-cultural determination of the functioning of the imagination's psychosomatic foundations leads to a loss of the imagination as an independent ability—to its dissolution into other human abilities."[3]

The inventory of the modern methodology of science and culturology has been supplemented by a new method and subject field of research called hermeneutics, which relies, in particular, on the imagination in teaching ways of understanding and interpreting texts as elements of spiritual culture.[4] The development of hermeneutics has led to the creation of the concept of a plurality of different types of understanding, which can be summed up into linguistic (learning the sense and meanings of expressions), psychological (learning the inner world of other people), and scientific-theoretical (knowledge of laws, causes, etc.) ones.[5]

The hermeneutic tradition in modern methodology allows identifying, on the one hand, the object of study that arises from the expansion of culturology over the field of psychosomatics and, on the other, the assimilation by psychosomatics of the system of socio-cultural factors as the sphere of culturology. G.I. Tsaregorodtsev and V.G. Erokhin, in bringing together various tendencies in research on the relationship between the psychic and the somatic (W. Cannon, I.P. Pavlov, S. Freud, H. Selye, K.M. Bykov, F. Alexander), have come to the conclusion that the scientific approach to the problem of psychosomatics involves a synthesis of P.V. Simonov's informational theory of emotions, P.K. Anokhin's theory of functional systems, and the biochemistry of brain hormones.[6]

This allows for bringing into unity separate study results on the mechanism of the effect of ideal images on the central nervous system, of the latter, in its turn, on the endocrine, and of the latter, in its turn, on the somatic of the organism. In such a union, there is a lack of a link that would explain the origin of ideal images themselves. This necessitates the addition to this approach of a conception that reveals the nature of images and connects them with the psychomotor, the behavioural, and the communicative level of human activity. Such a conception is the theory of practice's mediating role in the relationship between the naturo-material and the ideal.[7]

The category of practice (Praxis) allows, in particular, revealing the specifics of the norm and pathology in a person expressed as the concepts of health and illness, which, unlike the concepts of norm and pathology, bear a social burden.

In his *Theoretical Biology*, E.S. Bauer formulated the universal principle of biology as the principle of a stable imbalance of living systems. According to this principle, living systems are never in equilibrium and at the expense of their free energy constantly work against the equilibrium such as required by the laws of physics and chemistry.[8] With the appearance of humankind, this working-against is supplemented by both an external practical and an internal spiritual activity, which changes both the character of the negentropic processes and the criteria of the norm. In contrast to the biology of the animal world, the biology of the human being is pathological in nature since it is not adapted to function in non-social conditions. However, what is pathology for the biological system becomes the norm for the human being as a biosocial being.

The maintaining of the optimal functioning of the biosocial system constitutes precisely the content of the concept of health, often characterized in practice as the ability to perform labour operations, in addition to the optimal biological functioning.[9] Conversely, the failure to maintain that optimality is indicative of illness.

The view of the norm as a statistical average, as well as both a model and an ideal, has been superseded by it being interpreted as a functional biological optimum (V.P. Petlenko, A.A. Korolkov), that is, as the functioning of a living system that meets the criteria of efficiency and reliability. This more general interpretation of the norm includes the first two as particular cases, as well as, moreover, the interpretation of the norm as a negentropic process (V.F. Serzhantov).

In interpreting the norm and pathology from the position of thermodynamic parameters, biological criteria for distinguishing between these two concepts are lost, whereas the latter's reduction to the idea of biological optimality leads to the medical being reduced to the biological. Furthermore, the identification of the norm with optimality seems insufficiently correct methodology-wise since the phenomenon of pre-pathology, on this approach, automatically turns out to be outside the norm. Moreover, it is even intuitively clear that it is far from any deviation from optimality being ready evidence of pathology. All these above-noted methodological and theoretical difficulties demand a more rigorous analysis of the relationship between these two concepts.

To clarify it, not only the reality of the optimal functioning of a living system but also the possibility of achieving this optimality at the expense of the internal resources of the organism is the norm. Thus, the norm turns out to be a fluctuation around the optimal state and includes an interval between the optimal and the pathological called pre-pathology or the area of practical health.

D.N. Isaev, in considering the problem of psychosomatic interrelations, presents data according to which the more-than-double-the-average-number increase in the number of events leading to serious changes in a person's life leads to an up to 80 percent increase in the probability of illness. The author analyses a variety of stressors—psycho-traumatic events that can cause diseases that result from excessive stress in the course of a particular activity, anxiety over/about assessments/grades/evaluations or a disagreement in actions, or under the influence of physical or natural factors.[10] The main stressors currently include: "(1) fear of war and ecological disasters and anxiety over natural and industrial disasters; (2) suffering caused by national hostility and economic contrasts; (3) suffering caused by migration and loss of social status; (4) fear of attacks, harassments, and persecutions from others; (5) experiencing hardships associated with entering a school or other educational institutions, experiencing difficulties in learning a profession, and struggling in an atmosphere of rejection or hostility from fellow students or fellow employees; (6) mental traumatization due to being separated from loved ones; (7) experiencing wrong upbringing; (8) experiencing distorted intra-family communication/interaction; and (9) sufferings caused by antagonistic relationships among family members, lack of spiritual warmth, by harshness, sexual abuse, etc."[11]

In our view, various phenomena of spiritual culture are formed in such a way that they become able to compensate for or balance the traumatic psychogenic factors, such as those mentioned above. Thus the components of spiritual culture acquire the possibility of psychosomatic effects optimizing and normalizing the biosocial functioning of the human being.[12] Let us consider the impact of spiritual culture on the psychic and physical nature of the human being in various aspects: logical, gnoseological, axiological, psychological, and socio-historical.

The specification of the norm and pathology in psychosomatics occurs due to the inclusion of socio-cultural factors as the unity of material and spiritual culture, which are integral parts of human existence as being human proper. The norm in psychosomatics is qualitatively defined as health, i.e., as the maintaining of the optimality of functioning in its integrity, while quantitatively as the interval of such deviation from optimality as preserves the latter's integrity. When quantitative changes cross the boundary of the measure, making it impossible to maintain optimality, the system passes into a state of pathology. The inclusion in the norm of practical health explains the possibility of deviations or violations of particular elements of the psychosomatic mechanism when these are compensated for by other components, which preserves the norm for the entire system. For example, therapeutic intervention is possible in the course of psychosomatic processes at the levels of psychoanalysis, hypnosis, and teaching methods of self-suggestion and self-regulation in a state of trance. Different areas of alternative medicine ("extrasensory perception," "magic") have an impact on real psychosomatic processes.[13] Moreover, this impact has two peculiarities. The first consists in the difficulty of managing this process and in the problem of the ambiguity—unpredictability—of the result. The second peculiarity is the employment of ideological and religious metaphors borrowed from occult teachings and religions. Such images perform the functions of removing and shifting the focus of attention from the conceived pathogenic factor and replacing it with a different image, the consequences of the effects of which are uncertain. The presence of functional results from the employment of such metaphors does not yet indicate their gnoseological truth since they are phenomena of spiritual culture that carry the load of a symbol acquiring an individual associative meaning. This is the basis for the method of "magicians" and "psychics" expelling an uncertain anxiety or fear.[14] It consists of: (1) the transition from an indefinite, abstract to a specific, empirically concrete object or situation; (2) endowing such a concrete with magical qualities (by "techniques" of spell casting, hexing, removing the evil eye, etc.); (3) eliminating or overcoming the pathogenic object (or situation) in the memory; and (4) translating this [whole] process to the subconscious level as faith.

In the history of culture, the mass ousting of phobias caused by the view of the world as a source of danger was advanced by the emergence of the Christian God who treats the human being with a "fatherly" love lying in the sphere of pure spirituality as an abstract value of human existence. Spirituality as an ideal image, elevated to a cult in Christianity, had a significant impact on human psychology.[15]

Christianity as a new phenomenon of spiritual culture so influenced the psychology and inner world of the human being that it gave rise to a peculiar behaviour of people—a behaviour which allowed it to be regarded as the basis of the genesis of a specific ethnos. L.N. Gumilyov characterizes the emergence of Christians as the birth of a new super-ethnos formed during the first centuries of our era out of new people, who were unlike any of their neighbours and who already at the beginning of their appearance "opposed themselves to the 'pagans', that is, to all the rest, and indeed stood out from among them—not, of course, by anatomical or physiological features, but by the nature of their behaviour."[16]

In the psychological aspect, cultivating Christian love in the mind, transferring it to the sphere of the subconscious, and turning it into an object of faith contribute, among other factors, to ousting and transforming the instincts of the "libido" and the "thanatos,"[17] which neuroticize human life in society.

S. Freud's "theory of repression" became the centrepiece of his psychology. The application of the "repression mechanism"—a concept formed in the psychology of neuroses—to the analysis of dreams was one of the boldest attempts to solve the riddles of the unconscious psychic being on the ground of empiricism. C.G. Jung shows that "dream interpretation has become the key to the answers to many questions from the domain of normal psychology and the psychology of neurosis."[18] In particular, he [Freud] "opened the way to an understanding of the inner picture of the formation of hallucinations and delusions under schizophrenia, which had been previously described only as a combination of their external manifestations."[19] Jung also characterizes the socio-cultural origins of Freud's teachings, outlining the spiritual background necessary for the formation of a correct understanding of this scientist.[20]

In a scholarly debate with the European psychoanalytic school, Sri Aurobindo Ghose rightly points out that the immersion into the "hell" of the subconscious can be undertaken only on the basis of the superconscious ("supramental") as the bright principle in the human being. Only in this case would the human soul be ensured against possible traumatization.[21]

Of the ideal images that emerge in the history of spiritual culture, some such are cultivated which assist in the optimal functioning of the entire psychosomatic system of the human being—a functioning which accords with the practice and forms of social interaction. This makes it possible to identify the relations between the norms and values of culture and the norm in psychology, which gives rise to the possibility of revealing the more rigid conceptual foundations of such metaphors as "sick" society and "healthy" society.

The norm in psychosomatics serves as an integrative criterion of the norm in psychiatry, which practically follows four principles. These are the ability to concentrate attention on an object, to retain information in memory, to process it logically, and to adequately orientate oneself in space and time. Violations of some of the above-listed abilities can be compensated for by a maximization of others to ensure normal functioning. Pathology, however, implies a simultaneous violation of a number of abilities. In our view, it is possible essentially to clarify the concepts of norm and pathology in psychiatry by including in the criteria for their differentiation the ability to maintain optimal functioning—practical, professional. This is rather important because, from the standpoints of cultural norms fixed in ordinary consciousness, the behaviours and activities of people of genius or talent can be perceived as abnormal[22] and can even impact the process of professional diagnosis. Such a danger is especially great in totalitarian societies, which primitivize the norms of culture and social interaction.

The dialectic of the norm and pathology in psychiatry has a peculiarity in the sense that the fact of the norm presupposes pathological deviations in the course of the functioning of the individual components of the psyche as a whole. Moreover, the emergence of a norm of a higher order can be perceived as pathology from the standpoint of the old criteria.

Not confining himself to the psycho-physiological problematics in S. Freud, E. Fromm undertook to comprehend the psyche in unity with social factors, thereby broadening the criteria for the norm and pathology in psychiatry. Such a broader psychosomatic approach allowed him to outline the processes and tendencies of both the normal and the pathological development of society.

"The content of human history is, according to Fromm, constituted by the desire of people to find themselves—to realize those needs which are generated by the disintegration of the former, initially holistic, relations. However, a specifically established society, one or another form of social community, prevents the full realizations of human potentialities," such as "the human need for communication, interpersonal ties, creativity as one of the significant motivations of the human being, and a feeling of deep roots guaranteeing the strength and security of being, as well as the desire for assimilation and for the knowledge, mastery, of being."[23]

In his philosophical anthropology, which includes psychoanalytic, existentialist, anthropological, and Marxist ideas, as well as ancient mystical traditions, E. Fromm reveals the picture of the "universal alienation" of the human being in contemporary society. He shows that in relation to the individual the society can perform opposite functions: to promote the fulfilment of the potential abilities and needs of a person or to distort their inner incentives. In the latter case, there occurs the phenomenon of depersonalization, where the person loses contact with themselves and builds material relationships with other people on a functional basis. The alienated person, who believes that they dominate nature, becomes a slave to the objects and circumstances in the world of their own objectified powers, which deprive them of freedom and the most powerful of which are, according to Fromm, bureaucratization and informatization. Fromm believes that today's humanity waits with fear to see whether it will be saved or come under the burden of its own bureaucracy—whether it will be able to maintain its freedom in the information society, which in developing by the standards of technocratic thinking he believes threatens to stop being human.

In developing his characterology, Fromm introduces the concept of a social character as a stable and specifically defined system of orientations. The five different ways of socialization—masochism, sadism, destructiveness, conformism, and love—that he distinguishes correspond to their respective ways of social adaptation: receptive, exploitative, hoarding, marketing, and productive. Fromm puts this characterology at the basis of ethical research, the object of which becomes to a greater extent not the individual vices and virtues of a person, but their virtuous (productive) or vicious (unproductive) character.[24]

The historical and changeable character of the norm in psychosomatics is conditioned not only by the natural, but also by the socio-cultural environment. In the genesis and development of spiritual culture, there emerge as values and norms such ideal images as most closely correspond to the optimization of the psychosomatic functioning of the human being in various certain historical and regional conditions. Their psychosomatic effects depend on the impact on the human organism of words and other symbols that fall deep into consciousness and lead to physiological consequences. Thus, for example, verbal suggestion in a state of hypnosis can sharpen hearing, sight, and the sense of touch. Verbal exposure alone can affect heartbeat and blood circulation.[25]

In turning from the functional analysis of psychosomatic processes to the consideration of the physiological content of their main structural elements and the latter's connection with the material field of human activity, it is necessary to draw attention to the role of the neurodynamic models of the brain. On the one hand, these are a form of encoding information about the external world and, on the other, the organizing nucleus of all afferent and efferent information flows inside the organism—the centre where all somatic (neuromuscular, biochemical, hormonal, circulatory, etc.) processes converge.

The continuum of all these reactions, considered in the aspect of human social activity, both reflects and ensures human material activity. The corporeal and spiritual integrity of the human being "is formed," as G.I. Tsaregorodtsev and V.G. Erokhin note, "on the basis of the unity of two complementary phenomena—psychosomatic response and psychosomatic maintenance."[26] An analysis of the mechanism of psychosomatic relations indicates the absence of a sharply defined border between their structural and functional aspects. They turn out to be interrelated and complementary aspects of human life activity.

Moral and aesthetic cultural values and norms are recorded in the psychosomatic organization of the human being as the unity of signs and significations of various components of spiritual activity—components that are vital factors. Methods of their recording in spiritual culture are not confined to natural language and code systems derived from it, but include other forms of symbolization, such as music and visual arts. Thus, widely known is the effect used in medical practice of musical harmony and certain types of visual arts on the human psychosomatic system. However, the most powerful factor in psychosomatic affect is the word.

The foundation for comprehension of the fundamental role of the word in the development of spiritual culture is laid in the fundamentals of Christian doctrine. The statement from the Gospel of John "In the beginning was the word" characterizes the understanding of the human world from the standpoint of Christian spirituality. As the universal outlook of Christianity, it is also a prerequisite for cultivating the healing effect of the word on the psychosomatic nature of the human being. K.I. Platonov, a well-known psychiatrist, scientifically substantiated the role of the word as a psycho-physiological healing factor.[27] He examined a wide range of verbal effects on various physiological and even biochemical processes that occur in the human organism.

Currently, treatment with words ("logotherapy") has become an integral part of psychotherapy.

The word, which psychosomatically affects the human being, is a unity of sign and signification. Without analyzing these categories, it is impossible to reveal the genesis and development of spiritual culture. Language as the second signal system is formed during the process of the appearance and use of tools, which cause a sensual-visual image of a practical operation—an image that is superimposed on the sensual-visual image of a natural object. The result is an ideal signification of the image of a tool as a sign of transformation procedures, which are animized in the process of the development of spiritual culture and constitute the ideal in the gnoseological aspect.

The simplest stone tool is a sign signifying practical transformation (for example, splitting objects with the help of a chopper), which constitutes an ideal image as the signification of this sign. Therefore, in our view, it is possible to assume that a certain kind of cultivation of choppers historically took place, one that explains their production in quantities that exceed practical needs, as well as the emergence of "some kind of pre-aesthetic evaluations" in their production process.[28]

The mechanism of the genesis and development of spiritual culture has at its basis choice of a system of signs and significations, cultivated as a value in the specific conditions of social life activity. The fetishization of natural objects, which consists in transferring signification onto an external object as a sign, passes through the stages of animatism—the belief in the impersonal animacy of nature and its individual parts and animism—the belief in the spirits controlling people and natural operations.[29] These stages represent the initial stages of the formation of spiritual culture. Animism is introduced into mythology, epos, some esoteric teachings, and religion, which transforms it with its own instruments. The next stage in the self-cognition of spiritual culture can be philosophy, based on self-reflection and science, based on objectification.

The identification of sign and signification, attributed to the sign—their indivisibility—lies at the basis of not only fetishism, but also totemism, cults of various spheres of life activity, magic, shape-shifting, and quackery.

Fetishism is a form of the socio-cultural recording and development of sign and signification in the language of culture,[30] a prerequisite and method of recording socially significant values in the genesis of spiritual culture, and a prerequisite for writing and semiotics. So, for example, in the city of Sicyon Zeus was worshipped as a stone pyramid; a rough tree trunk on the island of Delos was referred to as the mother of Apollo and Artemis; in the city of Thespiae there was a sacred stone held to be the incarnation of the deity of love, Eros; in Sparta two logs connected by a crossbar were worshipped as two (twin) brothers, Castor and Pollux, sons of Zeus.[31] All these cases imply fetishes, in which dwells demonic power understood as the signification of these objects—signs.

A fetish is an object functioning as a sign to which the value of animacy is attributed. An animized image as a unity of sign and signification—a unity which explains why "myth is not an ideal being"[32]—can in particular be a fetish. In the fetish, sign and signification are combined in one object or image. The concept of cultivation is more general than fetishism and includes not only the phenomena of animism, but also self-reflexivity, ontologization, and objectification.

Thus, the phenomenon of fetishism is the initial nucleus of the analysis of spiritual culture, since fetish as a unity of sign and signification is a unit of socially significant value. In our view, it contains not only the rudiments of all religions, which was noted by D. Brinton,[33] but also of spiritual culture as a whole. At the level of the animistic awareness of spiritual culture, fetish, being by its nature the signification of the corresponding sign—that is, an image—itself is not yet recognized as an image. From the moment of it being comprehended as an image—that is, as a signification that does not coincide with the sign—fetishism is abolished by cultivating the ways of learning conventionality and literality in the construction of images.

A. Comte distinguishes between gods, in the proper sense of the word, with their general and abstract character, and pure fetishes (animate objects). While a modest fetish controls only one item with which it is combined, the gods simultaneously control a whole number of phenomena in different bodies. The well-known author of the animistic theory of primitive culture E. Tylor sees in this understanding a transition from fetishism to polytheism owing to the predominance of typal concepts over individual ones.[34] He observes that not all, but only the principal images or figures, such as Heaven and Earth, Rain and Thunder, Water and Sea, Fire and Sun and Moon, belong to Nature-worship in the strict sense of the word. These images or figures are either worshipped for themselves or are viewed as animized by special deities or are set apart as gods in anthropomorphic shape. All this yields a group of conceptions based entirely on savage fetishism. Tylor explains the ideas of the unreasonable power and unlimited influence of these great gods of nature by the fact that the objects of nature to which they refer are immense in size and strength and in remaining fetishes themselves dominate lesser fetishes. "In the religion of the North American Indians, the Heaven-god displays perfectly the gradual blending of the material sky itself with its personal deity."[35]

The improvement of cults consists in transforming the biological instincts of the subconscious and subordinating them to the progressively developed images of consciousness, which become an optimizing basis for interaction. In response to the needs of social change during the transition from primitive hunting to agriculture, zoomorphic gods are being displaced by anthropomorphic ones, with varying degrees of displacing old cults by new ones. By the period of the emergence of epic consciousness, this battle takes the form of an opposition between anthropomorphic gods endowed with the properties of good and evil, rational and non-rational principles, for example, Ormuzd and Ahriman in Zoroastrianism, Apollo and Zeus in ancient Greek mythology, and Krishna and the goddess Kali in the *Mahabharata* (the *Bhagavad Gita*). Zarathustra (Zarathushtra) in the *Avesta* speaks in the name of water, plant, animal, and human, justifying the good life of people by the good permeating the world itself and human behaviour corresponding to this good. In the *Avesta* (in the "Gathas") it is affirmed that it is necessary to replace hallucinogenic means of affecting a person with those of spiritual good in order to attain people's enlightenment through a meditation technique—that is, through psychological, spiritual, and not physical (somatic), means.[36]

In the *Rigveda* the psychogenic drink Soma is endowed with the properties of a spirit becoming an object of cult worship:

1. Thou, Soma, art preeminent for wisdom;	[You Soma, reveal yourself through our mind!]
along the straightest path thou art our leader.	[You lead (us) by the straightest path!]
Our wise forefathers by thy guidance, Indu,	[Under your guidance our fathers, O juice,]
dealt out among the Gods their share of treasure.	[Received a share in the wealth of the gods, (they are) wise]
2. Thou by thine insight art most wise, O Soma,	[You, by the strength of the spirit, are beautiful by the strength of your spirit,]
strong by thine energies and all possessing,	[You, by the power of action (effect), are beautiful by the power of your action (effect), omniscient,]
Mighty art thou by all thy powers and greatness,	[You, by the qualities of the bull, are a bull, by the power.]
by glories art thou glorious, guide of mortals.	[By refulgence, you have become refulgent, (you) with the gaze of a hero.]
3. Thine are King Varuna's eternal statutes,	[For you have King Varuna's statutes]
lofty and deep, O Soma, is thy glory.	[High and deep is your essence, O Soma.]
All-pure art thou like Mitra the beloved,	[You are as pure as the dear Mitra;]
adorable, like Aryaman, O Soma.	[You are like Aryaman who must be wisely worshipped.]
4. With all thy glories on the earth, in heaven,	[Those your forms which are in heaven, which are on the earth,]
on mountains, in the plants, and in the waters,—	[Which are in the mountains, in the plants, in the waters—]
With all of these, well-pleased and not in anger,	[In all these benevolent and not irritated,]
accept, O royal Soma, our oblations.	[O King Soma, accept our sacrificial oblations!]
5. Thou, Soma, art the Lord of heroes,	[You, Soma, are a true lord,]
King, yea, Vritra-slayer thou:	[You are also king, the slayer of Vritra,]
Thou art auspicious energy.	[You are the happy strength of the spirit.]
6. And, Soma, let it be thy wish	[If you, Soma, will want us]
that we may live and may not die:	[To live, we will not die.]
Praise-loving Lord of plants art thou.	[(You are) lover of praise, lord of trees.]

7. To him who keeps the law, both old and young,	[You, Soma, are a share for the old,]
thou givest happiness,	[You (also)—for the young righteous,]
And energy that he may live.	[You give the power of action (effect) (for them) to live.]
8. Guard us, King Soma, on all sides	[You, Soma, on all sides]
from him who threatens us:	[Guard us, O King, from the
never let	evil-intending!]
The friend of one like thee be harmed.[a]	[May the friend of one like you not suffer harm.]

[a]*The Rigveda*, Mandalas I–IV, trans. T.Ya. Elizarenkova (Moscow, 1989), 108–109, and *The Hymns of the Rigveda*, trans. Ralph T.H. Griffith, 2nd ed. (Kotagiri (Nilgiri), 1896), [01–091] HYMN XCI Soma, 48–49, web: http://www.sanskritweb.net/rigveda/griffith.pdf . See how my word-for-word translation from the Russian translation (in square brackets) compares with Griffith's.

In the text of the *Rigveda* "Soma" is used in its several meanings: (1) a substance from which juice is squeezed with the help of stones in a ritual to prepare amrita from it; (2) the squeezed juice; and (3) the God Soma, the bridegroom of Surya, the daughter of Sun.[37]

The next level in the development of spiritual culture, as exhibited in the *Rigveda*, is associated with the worship of spirituality itself, personified in gods who embody a moral function: Agni is the god of fire and the mediator between gods and people;[38] Bhaga, the giver of abundant wealth, is closely related to the weal-giving Ansha; Varuna is the all-inclusive god associated with the original world, guarding the cosmic law and punishing sinners; Mitra is a god who together with Varuna protects the universal law, and personifies the bright, auspicious origin of humanity.[39]

In the hymn "To Agni" there resounds a plea for deliverance "from all manifestations of hatred,"[40] a request for a benevolent and noble-hearted attitude of the gods and a hope for their protection—a hope that relieves of internal anxieties, worries, and human uncertainties and thus performs a psychotherapeutic function:

4. Even as true knower of the Law, O Agni,	[If only you, O Agni, would benevolently notice our]
to this our solemn rite be thou attentive.	[Effort (at the rite), order, O knower of order!]

When shall thy songs of festival be sung thee?	[When will the hymns sung under joint intoxication (by soma please) you?]
When is thy friendship shown within our dwelling?	[When will (your) friendly feelings dwell in (our) homes?]
5. Why this complaint to Varuna, O Agni?	[How can you (complain) on this to Varuna, O Agni,]
And why to Heaven? for what is our transgression?	[How can you complain to Heaven? What is our transgression?]
How wilt thou speak to Earth and bounteous Mitra?	[How will you to Mitra the generous, Earth, Speak?]
What wilt thou say to Aryaman and Bhaga?	[What to Aryaman? What to Bhaga?][a]

[a]*The Rigveda*, trans. T.Ya. Elizarenkova (Moscow, 1989), Mandalas I–IV, 362. My literal translation from the Russian translation in square brackets. English translation by Griffiths, p. 157.

Consideration of the history of cult development in the Vedic tradition attests to an increase in the spiritual component in cultivated gods, to the point of the personified deification of spirituality itself. Similar patterns can also be traced when analyzing the ancient Greek epos.

In Homer's poems, the general picture of the world can be viewed as the first—epic—model of the ancient cosmos: "it is like a huge dwelling in which people and gods live, act and fight."[41] Thus, in *The Iliad* Hera says to Aphrodite:

Give me the love and lust with which
You tame all immortal [gods] and mortal people,
For I am going to the extreme limits (πειρατα) of the food-bountiful Earth, to visit
Okeanos, the progenitor (γενετιν) of the gods, and mother Tethys.
It is they who kind-heartedly raised and groomed me in their home
Having received me from the hands of Rhaea when the loud-sounding Zeus
Had overthrown Chronos under the Earth and under the inexhaustible Sea.
I am going to visit them so as to end their endless quarrels (νειχεα).
For a long time now have they avoided each other

> And abstained from the [marital] bed and from love
> (φιλο΄τητοξ), for malice has come over their hearts.[42]

When this dwelling itself becomes alive and animate, something more authoritative than all the Olympian gods put together comes into existence for the ancient Greek reflecting upon the world. They would yet for a long time continue to remain an element of the Greek spiritual culture. "However, in essence, the traditional Greek polytheism was dealt a mortal blow on the very day that Thales of Miletus proclaimed water as the source and principle of all existence."[43] As for Homer, he focuses his attention on the description of the battle between the rational [intelligent] and the non-rational [unintelligent] principle in the human being. This battle between intelligence and impulse is depicted by Homer in the images of Hector and Achilles in his epic poem *The Iliad*.

As I.V. Shtal' and T.V. Popova note, intelligence is a "quality naturally inherent in every epic hero."[44] Hecuba addresses one of such heroes—the elder Priam with these words:

> Woe! Has your wisdom perished, for which in the past
> You have been famous both among foreign nations and in your own kingdom?
>
> (*The Iliad*, XXIV, 201–202)

Intelligence as a quality of epic heroes assumes a special role in *The Odyssey*. Telemachus is mostly intelligent. His main epithet is "intelligent." Menelaus is "very much intelligent." Odysseus is intelligent in the highest degree, "wiser than all men."[45] Penelope, besides having other virtues, stands out from all other women by her "intelligence," "prudence," and "intelligent shrewdness";[46] hence, the exclusivity of her wisdom, "having no decisive counterweight, naturally turns into some kind of absolute."[47] Without diminishing Odysseus' other qualities, which make him stand out as an epic hero from all the Denyen, Homer constantly emphasizes such virtues of his as "sagacity," "wits," "intelligent shrewdness," "ingenuity," and "intelligence":[48]

No one's sharp wit out there could ever openly rival
That of the god-like Odysseus….

(*The Odyssey*, III.120–121)

It is such qualities as "wits, cunning, ingenuity, and the most important virtues in *The Odyssey* that serve as protection from misfortunes, troubles, and sorrows, which fall to the lot of the human being in exceeding numbers, in violation of the epic harmony."[49]

The complication of the meanings of epic symbols, as contrasted with those purely mythological ones, forms the distinctive features of the epic—the unity of the conventionality and literalness of images,[50] including their increasing spirituality. The development in the epos of the three above-noted features—the literalness, conventionality, and spirituality of images—creates fundamentally new forms of spiritual culture.

This coincides with both the "discovery" of consciousness—a discovery prepared by ancient Greek lyric poetry, which turned to the description of the inner world of the human being[51]—and the overcoming of the historical limitations of the ordinary "unhappy consciousness"[52] in ancient (Greek) tragedy. A similar process can be seen in the development of the image of Orestes in *The Eumenides* of Aeschylus' tragic trilogy, *The Oresteia*. Persecuted by Erinyes but protected by Apollo, Orestes comes to trial to be judged—that is, he turns to judgement, which self-reflexive philosophical content helps [him] to overcome the inconsistency of his consciousness by displacing the latter's emotional oversaturation.[53] The option of going beyond the limited "unhappy" consciousness, i.e., the solution to Orestes' problem, is provided by the image of Antigone from Sophocles' tragedy bearing the same name. The heroine of the tragedy affirms the primacy of universal human value, which would later become the basis of Christian morality. Her positions are substantiated by means of reasoning in the course of the dialogues and as self-conscious come close to being philosophical while remaining characteristic of an artistic image.

Analysis of both the ancient Indian and the ancient Greek epos shows that with the emergence of generalizing cosmogonic models as the epic consciousness develops, the spiritual content of the principal images is enhanced. With awareness in the epos of the conventionality, spirituality, and literalness of mythological images, the prerequisites for art, religion, philosophy, and science are formed, respectively.

New knowledge systems based on the clarification of the literalness of images—that is, on revealing the content of the concept, on the conventionality of artistic images and their aesthetic construal, and on the spirituality of images—lead respectively to [the emergence of] philosophy, art, and religion. The psychotherapeutic effect of the first on the displacement of the emotional tension in consciousness is realized through reflection and understanding, that of the second through empathy, and that of the third through enlightenment (grace, mercy, and nobility).

In the formation of philosophy, the elements of previous perceptions that are sung in the epos are replaced by new—gnoseological—images—by such concepts as *"arche," "apeiron," "atoms,"* and *"eidos."* Replacing the elements of perception with concepts leads to the construction of a natural-philosophical ontological model—for example, the hierarchy of ideas in Plato, on top of which is the idea of the good.[54] "Plato had received from Socrates the great impulse, the active male principle of his life, his faith in justice and truth. [However,] [h]e was indebted for the science and substance of his ideas to his initiation into the [Eleusinian] Mysteries, and his genius consists in the new form, at once poetic and dialectic, he was enabled to give to them."[55] While Socrates as the creator of new knowledge is a philosopher, in the socio-cultural sense, he is the priest of the cult of the "new gods"—philosophical ideas.[56] The realization of the cult of these "new gods" takes place in the form of the famous Socratic dialogues.

In order to better understand the Platonic problem of the ideal, it must also be remembered that, originally, the problem of metaphysics developed as the idea of personal salvation, as was the case with, for example, the Pythagoreans. It is transcendence, "looking over and above one's own culture, one's own situation, etc., or, as the ancients said, detachment from the 'wheel of rebirths.'"[57] Transcending means going beyond one's own limits. But how can one do this through one's own thinking? "You can surely be in a state of transcendence; however, to gain a point of view and see what is transcendentally assumed is impossible. What does Plato do in this case? He moves up to the level of reflection. This is where the problem arises that we would much later on come across in epistemological garment in the works of Descartes, Locke, Kant, etc."[58]

The creative tension that emerges in the course of the philosophical comprehension of reality can have a psychosomatic effect. "A living thought makes a person more energetic, healthier, at once stronger and softer, less reserved, simpler and more open-minded, so that the joy of having a living thought as though spreads throughout the body and even affects some unconscious depths of the psyche."[59] Philosophy, in developing a living thought, contributes to broadening the limits of the norm in psychosomatics. "A living thought is enveloped by the breeze of singing joys, by an endless willingness to live and work, to be healthy and strong. A lively mind is charged with cheer and vigour."[60]

We can see that in its very genesis philosophy, in reflecting on culture, departs from it by going beyond it in content, while culture assimilates philosophy as an element, making its functioning subservient to its laws. The same goes for the relationships of art and religion with culture. In each of these new phenomena a holistic image of culture is created, which leads to the displacement of the cult by culture, just as the fetish was displaced by the cult.[61]

Replacing the elements of perception with representations leads to the creation of artistic images and types. Thus, J. Huizinga, in showing the process of the formation of poetic images, observes that "the Rigveda gives us a mystical paraphrase of primordial myth-matter at the hands of sacrificial priests, who have interpreted it ritualistically. The primordial Being, Purusha (i.e., man) has served as matter for the cosmos. All things have been formed from his body":

the moon came from his spirit,
the sun from his eye,
from his mouth came Indra and Agni,
from his breath the wind, from his navel
the sky, heaven from his head,
earth from his feet, and from his ears

the four quarters of the horizon;
thus did they (the gods) fashion the worlds.[62]

This example already contains general mechanisms for the formation of artistic images—mechanisms which consist in the fact that the elements of real perception are replaced by representations as conventional images. For example, the *Rigveda* develops a certain tradition of artistically designating sunrise as a scarlet cow. The word play is here reinforced by playing with mythological images become poetic: "morning dawns are at once scarlet cows."[63]

Besides the techniques of construing an artistic image as conventional, the ancient Indian tradition displays traces of both human and animal deification by gradually endowing each with corresponding spiritual qualities. Generally, the religious culture of India is characterized by a hierarchical construction of animized images. This hierarchy encompasses both nature and people.[64]

Upon animistic representations is built the hierarchy of the spiritual, which proves to be a sufficient means for comprehending the world in Indian philosophy, without resorting to the concept of the ideal. However, its confinement to the spiritual aspect inevitably leads to views involving the transmigration of souls, incarnation, with which the deification of animals and humans is associated. Overcoming the limitations of such views is possible only with the gnoseological analysis of human consciousness, which opposes the ideal to the animistic.

In art, animistic representations take the form of conventional images, which overcome animism in their own specific manner. The very conventionality of images, more pronounced and perceived in the European tradition, becomes the basis for the creation of artistic works. E. Tylor "finds a curious interest in transformation-myths like Ovid's, keeping up as they do vestiges of philosophy of archaic type."[65] Thus, for example, Daphne is turned into the laurel that Apollo honours in remembrance of her, the sorrowing sisters of Phaethon are changed into trees dropping blood and crying for mercy when their shoots are torn down. Such images were even employed in medieval poetry. "If these seem to us now conceits over quaint for beauty, we need not scruple to say so…. And if even the classic originals have become unpleasing, we need not perhaps reproach ourselves with decline of poetic taste. We have lost something, and the loss has spoiled our appreciation of many an old poetic theme, yet it is not always our sense of the beautiful that has dwindled, but the old animistic philosophy of nature that is gone from us, dissipating from such fancies their meaning, and with their meaning their loveliness."[66] These observations of E. Tylor contain a reference to the source of the formation of artistic images, on the basis of which a new—as contrasted to the epos—form of social consciousness is built—art.

Emotional reactions that occur in perceivers of art may participate in the optimization of the functions of the organism. "Aesthetically charged feelings transform, as it were, into motives for the psychic regulation of the physiological functions of the organism, contributing to their optimal coherence."[67]

From the standpoint of logico-gnoseological analysis, the displacement of the old cults of pagan religions by the speculative construct of pure spirituality—by the abstract value of human existence—is the object of a new cult in world religions. For example, to the model of the world thus constructed in Christian theology are attributed the features implicit in the Sermon on the Mount, which are the principles of the Christian theological conception.

Possible are both a rational comprehension of the origin of religious cults and the cultivation of rationality in both philosophy and science. Thus, the development of the metaphysical position in philosophy has passed through the following stages of answering the question of being: nature in antiquity, God in the Middle Ages, thinking and being in modern times, and the irrational needs of the human being in the non-classical philosophy of post-modern times. These major answers to the question of being determine the philosophical mentality of the epoch—a mentality that pervades the latter's spiritual culture. Similarly, a change occurs of image systems cultivated in science, which is reflected in the change of world pictures, paradigms, and fundamental theories, which serve as norms, values, and means of substantiating scientific knowledge.[68]

The content of the development of culture includes infusions of such cults as belonged to the earlier stages in human history and were overcome and displaced by more advanced cults. The real process of development, which included not only good, but also evil, not only the creation, but also the destruction of spiritual culture, was reflected and expressed by a generalized image symbolizing this destruction. The cult of the personified image of the powers of evil[69] rests in particular on the fact that in real life deviations from the ideals of morality and spirituality proposed by world religions constantly recur, which alongside culture contributes to the formation of counterculture.[70]

World religions, being a specific type of displacing of the early forms of animism (i.e., treating the good as an internal image), emerge in the context of aggravating conflicts between culture and civilization. Christianity arises in the conditions of the most acute conflict between the norms of universal morality as the quintessence of spiritual culture and the politico-legal structure of civilization. E. Renan showed that Christianity had originated when the communal-clan morality of Galilee—the most developed one in the universal sense—collided with the most developed—Roman—civilization, which displaced spirituality. The strong spiritual culture of the [politically] weak Galilee was a powerful factor in the fall of the weak-spirited Roman civilization.[71] In conditions in which human life meant nothing, salvation could [only] come from the cult of human life as the highest value; it is this value in the abstract that constitutes pure (bright, holy) spirituality as the main Christian cult.

Identifying the specifics of spiritual knowledge presupposes going beyond the limits of gnoseological toward axiological analysis. The gnoseological mechanism of the formation of the spiritual consists in idealization as a special kind of abstraction, which preserves some qualities (boundless mercy, love, grace) and detaches from others. The very content of such a construct is a moral-value attitude in the abstract toward the world and people, i.e., pure spirituality in the abstract. The transition from the new concept of God the Father to God the Son as a new idea is possible through eidetic interpretation (the Holy Spirit) and thought experiment (the Immaculate Conception). If these speculative constructs replace the concept of God in Judaism, then we get a new speculative model characteristic of Christianity.

The generalization procedure ascribes to this new speculative model relevant attributes, for example, those expressed in the statutes of the Sermon on the Mount. The utterances that ascribe attributes to the model act as principles that cohere with one another from the standpoint of logical non-contradiction. A system of such principles constitutes precisely the content of Christian theology. It rationally comprehends the axiological aspect of the worldview on the basis of the above-described speculative model, which within the framework of theology allows for turning the Socratic problem of reducing the good to knowledge into a reduction of knowledge to the good, of gnoseology to axiology, of truth to value, and of the ideal to the spiritual. The concept of the spiritual not only displaces the ideas of spirits, but also opposes them as animistic, just as the animistic is opposed by the ideal.

The psychological essence of Christianity is succinctly formulated by P.A. Florensky: "Religion is, or at least aims to be, a salvation artist, and its job is to save. What does religion save us from? It saves us from ourselves, it saves our inner world from its own hidden chaos. It overcomes the Gehenna within us—its tongues, which in breaking through the cracks of our souls, lick our mind. It strikes the reptiles of the 'great and vast sea' in our subconscious life—and 'they have no number'—and wounds the serpent nesting there. It settles the soul. And in bringing peace to the soul, it pacifies both all people and all nature.

That is the work of religion, taken deliberately within its most narrow and modest scope—its main work, which hardly anyone would want to dispute."[72]

In Christianity the formation of the rational comprehension of the relationships between the cult of pure spirituality (good) and the cult of destructive forces (evil) paralleled the development of the satanic cult.[73] These relationships are interpreted as between human and God on the one hand and human and devil on the other, and are recorded as respective legal norms in medieval Europe.[74]

The development of such forms of spiritual culture as philosophy, art, and religion—humanizing factors of society—helped to increase the level of the moral and intellectual health of humankind. At the same time, their opposition to one another in the history of culture led to painful traumatic experiences, widely reflected in artistic and historical literature. Resolving contradictions among different forms of spiritual culture and synthesizing various cultural phenomena based on self-reflection (philosophy), objectification (science), conventionality in construing images (art), and intuitive acceptance of spirituality (religion) would be possible at such a level of development when their own foundations were transformed in a manner that would allow for such integration.[75]

Philosophy as reflective thinking helps to trace the stages in the formation of thought in culture—thought which passes through the stages of animistic ideas, poetization, rhetoric, grammar, and logic. Philosophy, which includes both poetic and logical aspects (which is most clearly manifested in both its genesis in Plato and its developed modern state in M. Heidegger[76]), allows for not distinguishing between philosophical and philological aspects within the structure of its language.[77] Such a convergence of the poetic and the logical makes it possible to reveal both the polysemy of metaphorical ideas and the metaphorical content of polysemantic concepts as a logical and scientific characteristic of philosophical and scientific texts.

The two above-identified methods of studying different styles of thinking, which take into account the literal nature of conventional metaphorical images on the one hand and instances of metaphorism in strict, scientific concepts on the other, help to identify animization as the basis of both thinking in culture and philosophico-scientific non-classical rationality.

The ideal, being a product of reflection on the animistic content of thinking, differentiates within itself into logical and spiritual. Treating animistic images as literal constitutes the basis of magic, whereas taking them as conventional—which represents the initial stage of reflexivity, of comprehending them as images—constitutes the basis of art. Further idealization is associated with the polarization of conventional images into the spiritual (relation to content) and the rational (content of images). Treating the latter as literal forms the basis of scientific thinking, which is built not on animistic, but on ideal images, i.e., the non-conventionality of the scientific acts as a displacement of the literal nature of the animistic. The specificity of modern European science is determined by the subsequent displacement of the ideal in the objectified-material, which acts as the negation of the negation with respect to the animistic. Objectification, i.e., attributing the property of existence to ideal images, forms the relation to the world as to an image of the world. The latter constitutes the specifics of modern art, religion, philosophy, and science.

In post-Hegelian, non-classical philosophy, the subjective-objective opposition (as the world vs. the image of the world) is eliminated on the basis of interpreting the world as a representation. This is to be found not only in the role of practice (Praxis) mediating the subject-object relation, but also in a synthesis of being as representation and being as experience. Heidegger's raising of the question of being which includes the inquirer characterizes the specifics not only of non-classical philosophy, but also of relevant science and culture. This distinguishes non-classical from modern European philosophy, which correlates the world and the image of the world, as well as from ancient philosophy, which inquires into the "whatness" of being as the content of reflexivity.

The overcoming of the distinction between objective and subjective in non-classical philosophy renders the objectification procedure impossible (in the classical sense); and the latter is superseded by a procedure characterized in this study as animization, which, being the negation of the negation with respect to the animistic, is based on the ideal and overcomes the distinction between objective and subjective. Moreover, animization includes the spiritual not as an independently existing being, but as a relation to images indistinguishable in terms of their being conventional or non-conventional. The latter circumstance characterizes the peculiarity of such a phenomenon as "virtual reality" in contemporary technical science, based on computerization. In this case, animization yields a great variety of possible worlds, on the basis of which there unfolds the prospect of transforming the material world in the foreseeable development of human practice. Animizing thinking, which integrates various phenomena of spiritual culture, serves as the basis for the synthesis of philosophico-methodological and socio-cultural studies.

Non-classical philosophy, which is succinctly characterized by a metaphysical position toward being which includes the inquirer, expresses the specifics of non-classical science. The peculiarity of contemporary science manifests itself in quantum mechanics taking the position that the micro-object is dependent on the observer, which leads to the problem of hidden variables. Similarly, the anthropic principle in relativistic cosmology, which consists in the idea of the Universe having developed in the direction of the emergence of humankind, requires an adequate philosophical interpretation. The development of contemporary mathematics, which is associated with the emergence of the theory of fuzzy (uncertain) sets and the theory of topoi and which is a transition from the study of fixed sets to that of fuzzy procedural configurations, also does not fit into the paradigm of modern European thinking, which has being and thinking opposed to each other, in a subject-object relationship. Thus, the scientifico-methodological objectification procedure characteristic of the rationalism of classical science turns out to be insufficient for an adequate study of the features of contemporary non-classical science.

Notes

1. See N.S. Mudragei, *The Rational and the Irrational. A Historico-Theoretical Essay* (Moscow, 1985); G.V. Sorina, "Psychologism and Antipsychologism On the Role of Logic in Scientific Cognition (The End of the Nineteenth through the Beginning of the Twentieth Century)," *Philosophy of Science* 6 (1986); Gaston Bachelard, *New Rationalism* (Moscow, 1987); N.S. Avtonomova, *Understanding, Reason, and Rationality* (Moscow, 1988); I.T. Kasavin and E.A. Sokuler, *Rationality in Cognition and Praxis* (Moscow, 1989) and *Rationality As the Object of Philosophical Study* (Moscow, 1995); and P.P. Gaidenko, *Scientific Rationality and Philosophical Reason* (Moscow, 2008).
2. L.S. Korshunova and B.I. Pruzhinin, *Imagination and Rationality* (Moscow, 1989), 61.
3. Ibid., 12.
4. On the subject and method of hermeneutics, see Hans-Georg Gadamer, *Truth and Method. The Basics of Philosophical Hermeneutics* (Moscow, 1988) and *Hermeneutics: History and Modernity* (Moscow, 1985); E.K. Bystritsky and S.A. Kosharny, "Hermeneutics: The Gnoseological and Ontological Aspects of Critical Analysis," *Philosophy of Science* 1 (1987); P.P. Gaidenko, "From Historical Hermeneutics to the Hermeneutics of Being. Critical Analysis of the Evolution in M. Heidegger," *Questions of Philosophy* 10 (1987); I.G. Rebeschenkova, "The Interpretation of Natural Sciences in Hermeneutics (Critical Analysis of P. Hylan's

Conception)," *Philosophy of Science* 4 (1987); and V.G. Kuznetsov, *Hermeneutics and Humanistic Cognition* (Moscow, 1991).

5. V.V. Ilyin and E.A. Marinichev, "The Gnoseological Aspects of the Problem of Understanding in Natural-Scientific and Humanistic Knowledge," in *Natural Science and Socio-Humanistic Knowledge: The Methodological Aspects of Interaction* (Leningrad, 1990), 122.
6. See G.I. Tsaregorodtsev and V.G. Erokhin, *Dialectical Materialism and the Theoretical Principles of Medicine* (Moscow, 1986), 119–121.
7. See Karl Marx, "Theses on Feuerbach," in *K. Marx and F. Engels*, op., vol. 3 (Moscow, 1955–1974) and E.V. Ilyenkov, *Philosophy and Culture* (Moscow, 1991), 219.
8. See Ervin S. Bauer, *Theoretical Biology* (Moscow/Leningrad, 1935), 41–44.
9. On the role of the concept of optimality in determining the norm in medicine, see A.A. Korolkov and V.P. Petlenko, "Dialectics and the Theoretical Problems of Medicine," *Materialist Dialectics in Five Volumes. The Dialectics of Nature and Natural Science*, vol. 3 (Moscow, 1983), 310–316. On the optimality in practical activities, see O.S. Razumovsky, *Patterns of Optimization in Science and Practice* (Novosibirsk, 1990), 48.
10. See D.N. Isaev, "The Problem of Psychosomatic Interrelations," *The Methodology and Sociology of Pediatrics* (St. Petersburg, 1991), 117.
11. Ibid.
12. See A.E. Nazirov and O.V. Maslieva, "The Genesis of Spiritual Culture and Psychosomatics," *Medicine, Culture, Philosophy* (St. Petersburg, 1994).
13. See L.I. Vasilenko, "Magic: Old Evil or New Good?" in *Questions of Philosophy* 2 (Moscow, 1994) and Mircea Eliade, *Shamanism: Archaic Techniques of Ecstasy* (Princeton, NJ: Princeton University Press, 1972).
14. For the description of the nature of phobias, see S.S. Sizov, "On the Nature of Fear," *The Methodology and Sociology of Pediatrics* (St. Petersburg, 1991).
15. Josh McDowell refers to the following characterization of the psychotherapeutic effect of the Sermon on the Mount (see James T. Fisher and Lowell S. Hawley, *A Few Buttons Missing* (Philadelphia: Lippincott, 1951), 273): "If you were to take the sum total of all authoritative articles ever written by the most qualified of psychologists and psychiatrists on the subject of mental hygiene—if you were to combine them and refine them and cleave out the excess verbiage—if you were to take the whole of the meat and none of the parsley, and if you were to have these unadulterated bits of pure scientific knowledge concisely expressed by the most capable of living poets, you would have an awkward and incomplete summation of the Sermon on the Mount. And it would suffer immeasurably through comparison." (Josh McDowell, *More Than a Carpenter* (Moscow, 1992), 33).
16. L.N. Gumilyov, *Ethnogenesis and the Biosphere of the Earth* (Leningrad, 1990), 113.
17. See Sigmund Freud, *Psychology of the Unconscious* (Moscow, 1989), 382–439.
18. Carl Gustav Jung, *The Phenomenon of the Spirit in Art and Science* (Moscow, 1992), 72.

19. Ibid. Or: In particular, he "opened the way to an understanding of schizophrenic hallucinations and delusions from the inside, whereas hitherto psychiatrists had been able to describe them only from the outside." English translation of the quote from Carl Gustav Jung, *Spirit in Man, Art and Literature*, in *The Collected Works of C.G. Jung*, vol. 15, *Bollingen Series XX*, ed. and trans. Gerhard Adler and R.F.C. Hull (Princeton, NJ: Princeton University Press, 1971), 64.
20. "All that glittered in the nineteenth century was very far from being gold, religion included. Freud was a great destroyer, but the turn of the century offered so many opportunities for debunking that even Nietzsche was not enough. Freud completed the task, very thoroughly indeed. He aroused a wholesome mistrust in people and thereby sharpened their sense of real values. All that gush about man's innate goodness, which had addled so many brains after the dogma of original sin was no longer understood, was blown to the winds by Freud, and the little that remains will, let us hope, be driven out for good and all by the barbarism of the twentieth century. Freud was no prophet, but he is a prophetic figure. Like Nietzsche, he overthrew the gigantic idols of our day, and it remains to be seen whether our highest values are so real that their glitter is not extinguished [in the Acherontian flood]." (Carl Gustav Jung, *The Phenomenon of the Spirit in Art and Science* (Moscow, 1992), 74–75. English translation of the quote from Carl Gustav Jung, *Spirit in Man, Art and Literature*, in *The Collected Works of C.G. Jung*, vol. 15. *Bollingen Series XX*, ed. and trans. Gerhard Adler and R.F.C. Hull (Princeton, NJ: Princeton University Press, 1971), 66.
21. "The forces that stand in the path of the *sadhana* are the forces of the lower mental, vital and physical nature. Behind them stand the hostile forces of the mental, the vital and the subtle physical world. These can be dealt with only after the mind and heart become concentrated and focused in a single striving towards the Divine" (Sri Aurobindo Ghose, *Basics of Yoga* (St. Petersburg, 1993), 4).
22. See further N.E. Goncharenko, *The Genius in Art and Science* (Moscow, 1991), 351–369. Hegel gives an interesting example involving Newton, who through absent-mindedness "once grabbed a lady's finger to stuff tobacco into his pipe with its help" (G.W.F. Hegel, *The Encyclopedia of the Philosophical Sciences*, vol. 3, in *Philosophy of the Spirit* (Moscow, 1977), 189–190).
23. P.S. Gurevich, "The Prominent Thinker of the Twentieth Century," in Erich Fromm, *The Heart of Man* (Moscow, 1992), 10–11.
24. See Erich Fromm, *The Heart of Man* (Moscow, 1992).
25. As L. Ron Hubbard writes: "by knowing function alone, we can cure aberrations and psychosomatic ills any time, predict new ills and conditions, and generally, "work miracles," as such actions were once called before Man knew anything about the mind" (Lafayette Ronald Hubbard, *Dianetics: The Modern Science of Mental Health* (Moscow, 1993), 122. Translator's note: English quote from the 1950 edition brought by the Ron's Org Grenchen, Switzerland, p. 87, web: https://ronsorg.ch/wp-content/uploads/Dianetics-The-Modern-Science-of-Mental-Health.

pdf). "Aberration" is "a departure from rational thinking or behavior that occurs mainly because of 'engrams'" (ibid., 547). According to Hubbard, an engram is a recording in the human psyche of "the moment of pain and unconsciousness and the actual or apparent threat to survival" (ibid., 551).

26. There are two main groups of problems in the field of psychosomatic medicine. "The first group is the study of 'vertical' relations between the soul (the psyche) and the body after the 'top-down' principle. In its most general form, it is the problem of transcoding social information into the structure of behaviour, as well as into the psycho-emotional and morpho-physiological maintaining of the latter.... The second group of problems is concerned with the study of the 'bottom-up' relations of the somatic and the psychic." Here, in particular, is considered the "direct somatic determination of psychic disorders or, on the contrary, of a stable psychosomatic state socially experienced by a person as health and well-being" (G.I. Tsaregorodtsev and V.G. Erokhin, *Dialectical Materialism and the Theoretical Principles of Medicine* (Moscow, 1986), 122).
27. See K.I. Platonov, *The Word as a Physiological and Healing Factor* (Moscow, 1957).
28. A.D. Stolyar, "K. Marx's Thesis on the Material 'production of ideas' in Light of Archaeological Sources," in *The Soviet Archeology* 4 (Moscow, 1983), 27. The proposed explanation of the early cult of labour tools is indirectly confirmed by archaeological facts that reveal ancient burials with stone tools. The ancient customs of making gifts and sacrifices to the dead were considered by H. Spencer, E. Tylor, and others. As S.A. Tokarev notes, an understanding of the origin of such rites was clearly formulated by L.Ya. Shternberg: "If the afterlife is a continuation of the earth life, then from here naturally arose the idea that the deceased should be provided with everything that was necessary for them to exist in this world. Therefore, they are given clothes, vehicles, tools, weapons, etc." (S.A. Tokarev, *Early Forms of Religion* (Moscow, 1992), 169). It is logical to assume that the process of gradually endowing the stone with magical qualities starts with the manufacture of stone tools—a process widespread in the teachings of practical magic and alchemy and recorded, in particular, in the legends of the Orient (see *The Cosmic Legends of the East*, authored and compiled by S.V. Stulginskis (Moscow, 1992), 136–141). It can be posited that the mystical attitude to stones contributed to gold and silver being set apart to become a universal equivalent in the exchange of goods. "These objects, gold and silver, just as they come out of the bowels of the earth, are forthwith the direct incarnation of all human labour. Hence the magic of money." (Karl Marx, *Capital*, vol. 1, in *K. Marx and F. Engels*, op., vol. 23 (Moscow, 1955–1974), 102. English translation of the quote from Karl Marx, *Capital: A Critique of Political Economy*, vol. 1, trans. Samuel Moore and Edward Aveling, ed. Frederik Engels (Moscow: Progress Publishers, 1965), 63–64, web: https://www.marxists.org/archive/marx/works/download/pdf/Capital-Volume-I.pdf). With the

development of (various) forms of fetishism, under conditions where "the behaviour of men in the social process of production is purely atomic" (ibid., 102–103; translator's note: see English ed., 64), their interrelation (behaviour) assumes a material character primarily manifested in the fact that the products of labour take the form of commodities (ibid., 103; translator's note: see English ed., 64). "Hence the riddle presented by money is but the riddle presented by commodities; only it now strikes us in its most glaring form" (cf. my translation from the Russian translation: "Hence the riddle of the monetary fetish is but the eye-blinding commodity fetish now become visible") (ibid.).

29. See Lucien Lévi-Bruhl, *Primitive Mentality* (Moscow, 1930); James George Frazer, *The Golden Bough. The Study of Magic and Religion* (Moscow, 1980); Edward Burnett Tylor, *Primitive Culture* (Moscow, 1989); V.G. Panov, *Emotions, Myths, Mind* (Moscow, 1992); and S.A. Tokarev, *Early Forms of Religion* (Moscow, 1992).
30. On the sign specificity of the language of culture, see M.K. Petrov, *Language, Sign, Culture* (Moscow, 1991).
31. See A.A. Takho-Godi, *Greek Mythology* (Moscow, 1989), 27.
32. A.F. Losev, *Philosophy, Mythology, Culture* (Moscow, 1991), 25.
33. Daniel Garrison Brinton, *Religions of Primitive Peoples* (New York/London: G.P. Putnam's Sons, 1897), 134.
34. See Edward Burnett Tylor, *Primitive Culture* (Moscow, 1989), 386–387, and vol. 2 of the English edition in two volumes (London: John Murray, Albemarle Street, 1871), 231.
35. Ibid.
36. See *The Avesta*, compiled by I.V. Rak (St. Petersburg, 1997), 145–146.
37. See T.Ya. Elizarenkova, "A Dictionary of Major Mythological Characters and Ritual Concepts," in *The Rigveda*, trans. T.Ya. Elizarenkova (Moscow, 1989), Mandalas I–IV, 761–762.
38. See, for example,

> 1. THEE Agni, have the Gods, ever of one accord,
> sent hither down, a God, appointed messenger,
> yea, with their wisdom sent thee down.
>
> (*The Hymns of the Rigveda*, trans. Ralph T.H. Griffith, 2nd ed. (Kotagiri (Nilgiri), 1896), Mandala 4, p. 155)

Cf. My translation from the Russian translation:

> For you, O Agni, have the unanimous gods—
> A God has always been appointed master (of rite)—
> By the force of the spirit alone appointed.
>
> (*Rigveda*, trans. T.Ya. Elizarenkova (Moscow, 1989), Mandalas I–IV, 357)

39. See T.Ya. Elizarenkova, "Notes," in *The Rigveda*, trans. T.Ya. Elizarenkova (Moscow, 1989), Mandalas I–IV, 726, and "A Dictionary of Major Mythological Characters and Ritual Concepts," ibid., 758–760.
40. *The Rigveda*, trans. T.Ya. Elizarenkova (Moscow, 1989), Mandalas I–IV, 357. Cf. "all those who hate us" (*The Hymns of the Rigveda*, trans. Ralph T.H. Griffith, 2nd ed. (Kotagiri (Nilgiri), 1896), 155).
41. I.D. Rozhansky, *The Development of Natural Science in Antiquity* (Moscow, 1979), 124. See also F.Kh. Kessidy, *From Myth to Logos* (Moscow, 1972), 91–94.
42. *Fragments by Early Greek Philosophers*, part 1, compiled by A.V. Lebedev (Moscow, 1989), 33. My literal translation from the Russian translation.
43. I.D. Rozhansky, *The Development of Natural Science in Antiquity* (Moscow, 1979), 124.
44. I.V. Shtal' and T.V. Popova, "On Some Techniques of Constructing an Artistic Image in Homer's Poems and Byzantine Epos of the Twelfth Century," in *Antiquity and Byzantium* (Moscow, 1975), 59.
45. Ibid., 71.
46. Ibid., 71–72.
47. Ibid., 72.
48. Ibid., 72–73.
49. Ibid., 72–73.
50. See F.Kh. Kessidy, *From Myth to Logos* (Moscow, 1972), 67–81; V.V. Kim and S.Yu. Milyaeva, "Science and Extra-Scientific Forms of Rationality (Toward the Problem of the Genesis of Rationality)," in *Rationality in Science and Practice: Patterns of Convergence* (Sverdlovsk, 1989), 10–11.
51. See O.V. Maslieva, "The Origins of Lyrical Poetry (Artistic Consciousness in Ancient Greece)," in *The Contexts of Culture: Anthropological and Axiological Aspects* (St. Petersburg, 2008).
52. G.W.F. Hegel, *The Phenomenology of Spirit* (St. Petersburg, 1992), 112.
53. See A.V. Akhutin, "The Discovery of Consciousness (Ancient Greek Tragedy)," in *Man and Culture* (Moscow, 1990), 28–30.
54. It is possible to suggest the inter-relationship of the contents of the concepts of "Good," "Bhaga," and "God" [Russian: *Blago, Bhaga, Bog*].
55. Édouard Schuré, *The Great Initiates: An Essay on the Esotericism of Religions* (Moscow, reprint edition, 1914), 320. Translator's note: See *The Great Initiates: A Study of the Secret History of Religions*, trans. Gloria Rasberry (Hudson, NY: Steinerbooks, 1989); quotation translated by F. Rothwell in Édouard Schuré, *Hermes and Plato* (London: William Rider & Son, 1909), 83.
56. See Plato, *The Apology of Socrates*, op. in 3 vols., vol.1 (Moscow, 1963).
57. M.K. Mamardashvili, "Consciousness as a Philosophical Problem," *Questions of Philosophy* 10 (1990): 11.
58. Ibid., 12. See also M.K. Mamardashvili, *As I Understand Philosophy* (Moscow, 1992), 41–56, and N. Lobkovicz, "From Substance to Reflection. The Way of Western European Metaphysics," *Questions of Philosophy* 1 (1995).

59. A.F. Losev, *The Daring Spirit* (Moscow, 1989), 5.
60. Ibid., 6.
61. The Latin "*cultura*" has a number of meanings: direct—"cultivation, sowing, tending"—and metaphorical—"teaching/learning, education; worship" (see Ananyev, Yasnetsky, and Lebedinsky, *An Abridged Latin Dictionary*, (Moscow: P.M. Leontyev, 1883), 286). The Russian term *kul'tura* ("culture") means "cultivation and tending, cultivating [and] tillage" and "education, both intellectual and moral" (see V.I. Dal', *The Explanatory Dictionary of the Living Great Russian Language*, vol. 2 (Moscow, 1989), 217).
62. Johan Huizinga, "*Homo Ludens*. Study Results of the Play-Element in Culture," in *The Self-Consciousness of European Culture In the Twentieth Century* (Moscow, 1991), 84. (See also Johan Huizinga, *Homo Ludens* (Moscow, 1992), 138–167, 179–196. Translator's note: Translation of the quote from Johan Huizinga, *Homo Ludens: A Study of the Play-Element in Culture* (London: Routledge & Kegan Paul, 1949), 136–137.) The idea of living cosmos as an archaic element of consciousness passes through the whole of the history of spiritual culture, up until modern times (see Mircea Eliade, [*The Myth of the Eternal Return*: *Or*,] *Cosmos and History* (Moscow, 1987); U.A. Radzhabov, "From Myths to Modern Cosmological Conceptions," *Philosophy of Science* 7 (1991); and V.P. Filatov, "The Living Cosmos: The Human Between the Powers of Earth and Heaven," *Questions of Philosophy* 2 (1994)).
63. T.Ya. Elizarenkova, "Notes," in *The Rigveda*, trans. T.Ya. Elizarenkova (Moscow, 1989), Mandalas I–IV, 723.
64. Thus, for example, Sri Chaitanya's teaching identifies two categories of living entities: the eternally liberated and the eternally conditioned. The latter can be divided into two types: the non-moving (e.g., trees) and the moving (e.g., birds and beasts). The moving entities—"*jangama*" (Sri Srimad A.Ch. Bhaktivedanta Swami Prabhupada, *The Teachings of Sri Chaitanya* (Vilnius-Moscow-Leningrad-Calcutta-Bombay-New Delhi, 1989), 56. Translator's note: See English ed.: A.C. Bhaktivedanta Swami, *The Teachings of Lord Chaitanya* (New York, 1968), 52, web: http://www.prabhupada-books.de/pdf/Teachings-of-Lord-Chaitanya-1968.pdf)—are divided, in turn, into three subtypes: those flying in the sky, those swimming on the water and those walking on land. People make up only a small part of living entities on dry land. "Out of that small quantity of human beings there are many who are actually without any advancement in spiritual science, and are without cleanliness in their habits, without any faith in the existence of the Supreme Personality of Godhead" (ibid., Translator's note: See English ed.: ibid., 53). According to this teaching, such people "are simply like animals" (ibid.). The believers in the values of the sacred scriptures and spiritual development are called "*Aryans*" (ibid.). They fall into two classes: the righteous and the unrighteous. Of the many righteous who are engaged in "fruitful [karmic] activities" (ibid., 57. Translator's note: See English ed.: ibid.) for sense gratification only a few come to know the Absolute Truth. These are philosophers—empirics—called "*jnanis*"

(ibid.). Of many hundreds and thousands of philosophers, only a few achieve liberation. "This liberation means that theoretically one can understand that a living entity is not made of any of the material elements; that he is spirit soul, different from matter. Simply by understanding this doctrine, even theoretically, one may be called mukta, or liberated; but actually a mukta, or liberated soul, is he who understands his constitutional position as part and parcel eternal servitor of the Lord. Such liberated souls, if they are engaged with faith and devotion in the service of the Lord, are called Krishna-bhaktas, or Krishna conscious persons." (Ibid.)

65. Edward Burnett Tylor, *Primitive Culture* (Moscow, 1989), 374. Translator's note: See English ed.: Edward Burnett Tylor, *Primitive Culture,* in two volumes, vol. 2 (London: John Murray, Albemarle Street, 1871), 220.
66. Ibid., 374–375. Translator's note: See English ed.: 220–221.
67. V.P. Petlenko, V.I. Pogorelov, and A.S. Popov, "Medicine and Aesthetics. The Aesthetic Norms in Medicine," in *Historical Materialism and Medicine* (Leningrad, 1981), 178.
68. See the works by V.P. Bransky, M.V. Mostepanenko, A.M. Mostepanenko, V.S. Stepin, Thomas Kuhn, and Imre Lakatos.
69. See E.I. Parnov, *The Throne of Lucifer*: *Critical Essays on Magic and Occultism* (Moscow, 1991).
70. In creating fetishes contrary to established values, counterculture realizes its internal opposition to the latter, using, in particular, the traditions of esoteric teachings. (See *Encyclopedic Dictionary of Philosophy*, eds. L.F. Ilyichev, P.N. Fedoseev, S.M. Kovalev, and V.G. Panov (Moscow: Sov. Entsiklopedia, 1983), 275. E.I. Parnov, in describing the positions of contemporary subverters of spiritual culture, provides examples of the unnatural symbiosis of culture and insanity, which, in their opinion, "sanctify" each other, and shows what can "ultimately schizophrenize society, turning the art of the word into the field of action of the irrational" (E.I. Parnov, *The Throne of Lucifer* (Moscow, 1991), 13). The features of schizophrenic thinking (mentality) were researched by the well-known Russian psychologist L.S. Vygotsky, who, in going through [a series of] experiments, found that "the most significant feature of the schizophrenic psyche is a violation of the function of the formation of concepts" (L.S. Vygotsky, "Problems of Child Mental Development," in *Selected Psychological Studies* (Moscow, 1966), 482). What happens is "the splitting of that psychological system which lies at the basis of thinking in concepts" (ibid.), the disintegration and destruction of the function of the formation of concepts (ibid., 482–483), which are displaced by "complexes" in which "there is a connection between [some] individual links, but no connection among all the elements of the complex as a whole" (ibid., 484).
71. See Ernest Renan, *The Life of Jesus* (Moscow, 1991), 91–103.
72. P.A. Florensky, "Opening Remarks at the Master's Degree Defense of the Book *On Spiritual Truth*. Moscow, 1912. Delivered on the 19th of May 1914," in P.A. Florensky, *The Pillar and Ground of the Truth* (II) (Moscow, 1991), 818.

73. See Yu.A. Sandulov, "Christianity and the Cult of the Devil," in *The History of Religion* (St. Petersburg, 1997), 241–268.
74. See Jacob Sprenger and Henricus Institoris, *The Hammer of Witches* (Moscow, 1990).
75. Attempts at a synthesis of philosophy, science and religion have been undertaken by E.P. Blavatskaya, V.I. Nesmelov, and V.S. Solovyev.On the problem of the synthesis of science, philosophy and religion, see Yu.S. Vladimirov, *Basics of Physics, Philosophy and Religion* (Kostroma: The International Innovative Research Centre for Alternative Educational and Social Technologies, 1996); L.A. Markova, "On the Possibility of the Correlation of Science and Religion," *Questions of Philosophy* 11 (1997); and Yu.I. Kulakov, "The Synthesis of Science and Religion," *Questions of Philosophy* 2 (2000).
76. See Martin Heidegger, *Works and Reflections of Different Years* (Moscow, 1993).
77. See Jacques Derrida, "Spurs: Nietzsche's Styles," in *Philosophy of Science* 2 and 3 (Moscow, 1991).

Chapter Two

The Attempt at a Pre-theoretical Synthesis of Knowledge in Theosophy

The term "Theosophy" is derived from the combination of the Greek terms "Theos," God, and "Sophia," Wisdom. Originally, theosophy appeared to be identical to theology, whereas later it "came to be called a body of teachings about the deity that proceed from subjective mystical experience and seek to expound this experience as a coherent system."[1] Such a broad understanding of theosophy makes it possible to relegate Gnosticism, Neoplatonism, Kabbalah, and occult teachings to theosophy. However, this term was more widely recorded in sixteenth–eighteenth century mystical teachings, which stood outside of the Christian church tradition and which were represented by such names as J. Boehme, Paracelsus, L.C. de Saint-Martin, E. Swedenborg, and F. Oetinger. In the nineteenth century, there develops a tradition of understanding theosophy as a synthesis of the mystical knowledge of God and rational philosophy.

Thus, theosophy forms into an independent tendency standing out from the esoteric and occult teachings; however, it continues to be in close contact with the latter. Theosophy as an independent tendency takes shape in the nineteenth century, whereas occultism, according to researchers, comes from ancient Egyptian and ancient Indian teachings and its traditions divide into two branches: Western, of Egyptian origin, and Eastern, or Indian, one which was previously cultivated in India and which later took the form of theosophism in Europe. The legends of the Western branch rest on Hermes, Moses, Orpheus, Homer, Pythagoras, and Plato. The sect of Essenes kept alive the oral tradition of the secrets hidden in the books of Moses, and this tradition became known as Kabbalah. In the Middle Ages, occultism is represented by alchemists, astrologers, and the whole Order of Templars.

At the end of the eighteenth century, and in the nineteenth century, works on occultism are published more openly and its representatives are not persecuted by the church. This area of literature includes works by Martinez de Pasqually, [Louis] Claude de Saint-Martin, Vronsky, [Antoine] Fabre d'Olivet, Luis Luc, Laguria, Éliphas Lévi, Stanislas De Guaita, Joséphin Péladan, Papus, [Alexandre Saint-]Yves d'Alveydre, [Paul] Sédir, Ernest Bosc, Mario Decresp, and [François-Charles] Barlet. It is known that Stanislas Guaita [and Joséphin Péladan] founded the Kabbalistic Order of the Rosicrucians [the Rose-Croix, or the Rosy Cross], and Papus the "Independent Esoteric Study Group."[2] Occultism first of all draws attention to the facts inexplicable from the viewpoint of exact science and synthesizes them with the peculiarities of the human psyche. Thus spiritism, magnetism, hypnotism, telepathy, clairvoyance, etc. fall into the field of vision of the occult.

Occultism, in studying the psychic world of the human, aims to overcome the material aspect of consciousness, but it does not turn, as it is done in the humanities, to its ideal aspect. Therefore, in introducing the concept of the soul, the occultists remain in their [style of] thinking at the pre-philosophical level, namely, at the level of animistic thinking. The latter circumstance allows for understanding why the occult-spiritual teachings belong to the field of hypotheses that are outside of the limits of philosophy and science.[3]

Since in occultism signs and symbols are widely used to mediate logical, universal-conceptual thinking, their meaning is accessible only to the initiated, while treating it as something mysterious turns occultism into esotericism. The fusion of various styles of thinking (animistic, poetic, and sign-semantic) in occultism both strengthens and weakens it. Its strong side was manifested in the attempt to synthesize philosophy, science, and religion, which was undertaken in particular by E.P. Blavatskaya in her theosophy, and its weak one in the absence of the possibility of verification and falsifiability of such knowledge. Historically, in modern European thinking, the Cartesian "cogito" program suppressed occult teachings; however, with the "collapse" of classical rationalism (in the second half of the nineteenth century), the occult-spiritual tradition actively resurfaced in culture in the form of theosophy and anthroposophy.

E.P. Blavatskaya's theosophy, "in discarding the 'historical forms of religion,'" aimed to bring into unity various religions by revealing the identity of the sacred meaning of all religious symbols and to create on this basis a kind of "universal religion" not bound by any particular dogmatism."[4] To this we can add that theosophy originally claims to synthesize not only religions, but also philosophy and science. Taking into account the global nature of the problem posed and the need of contemporary spiritual culture for a synthesis of its main phenomena—a need created by the formation of a novel thinking at the turn of the third millennium, E.P. Blavatskaya's works[5] are of interest for both logico-gnoseological and philosophico-methodological analysis.

Hegel observed in referring to the knowledge contained in the ancient Indian Vedas as the source of the philosophy and religion of India that the latter two are essentially identical with each other. "Indian Philosophy… stands [moves] within [the framework of] Religion just as scholastic Philosophy stands [moves] within [the framework of] Christian dogmatism, having at its basis and presupposing the doctrines of the church [the Christian faith as its prerequisite]."[6] Indian philosophy, unlike European, includes features that substantially bring it closer to mythology and religion. This concerns, for example, the idea of one universal substance out of which everything takes its origin, and the summit of human consciousness is its identification with this substance. Moreover, at the same time, such a merger is achieved "[i]n Religion by means of worship [piety], offerings, and rigid acts of expiation [fasting], and in Philosophy through the instrumentality of [deepening into] pure thought."[7]

According to Hegel, Eastern peoples possess individual images, such as Brahma, Vishnu, and Shiva, but due to lack of freedom, their individuality is not clearly outlined. Therefore, the images that European thinking perceives as individual spread and grow vague in the Oriental worldview.

It can be posited that these worldview particulars of general understanding, which allow Hegel to call them "philosophemes," are conditioned by the specific quality of the organization of social life. Tribal relations, which were preserved in Eastern civilizations to a greater extent than in [Western] antiquity, did not make an individual free from the power of the community. Ways of influencing and controlling the individual in the community are reinforced by traditions and customs that are animized in taking the form of cults. At the same time, the hierarchy of animized images is conditioned and reinforced by the hierarchy in the management of communal life in the conditions of Eastern civilizations, which are based on the Asian mode of production.

Hegel showed that the point of main importance in Indian philosophy is the "becoming explicit [becoming for itself] of soul in the most abstract mode"; this "may be called intellectual substantiality; but here it is not the unity of mind [spirit] and nature that is present, but directly the opposite."[8]

Intellectual substantiality is an aim and an end for Indian philosophy, whereas, for the European, it is the beginning of philosophizing, the opposite of reflection. Thus, intellectual substantiality in Indian philosophy is opposed to the particular and does not become a moment in the self-developing cognitive system of thinking. To put it otherwise, it itself does not become an object of reflexive cognition, which would subvert it in its material truth. This lack of reflexivity determines the specifics of Eastern philosophy, as well as of the esoteric teachings based on it—specifics which consists in constructing a hierarchical system of animized images. Let us consider this by using as an example the teaching expounded in E.P. Blavatskaya's *The Secret Doctrine*.[9] The book sets forth the following main principles that ground the "Secret Doctrine:"

> (1.) The ABSOLUTE; the *Parabrahm* of the Vedantins or the one Reality, Sat, which is, as Hegel says, both Absolute Being and Non-Being. (2.) [*The first*

> *Logos*:] The first manifestation, the impersonal, and, in philosophy, *unmanifested* Logos, the precursor of the 'manifested.' This is the 'First Cause,' the 'Unconscious' of European Pantheists. (3.) Spirit-matter, Life; the 'Spirit of the Universe,' the Purusha and Prakriti, or the *second* Logos. (4.) [*The third Logos*:] Cosmic Ideation [thought-basis], Mahat or Intelligence, the Universal World-Soul; the Cosmic Noumenon of Matter, the basis of the intelligent operations [rational manifestations] in and of Nature, also called Maha-Buddhi.[10]

The listed grounds of the "Secret Doctrine" evince, to use the language of European philosophy, some kind of unity among the material, the spiritual, and the ideal. It can be posited that the unity of the spiritual and the ideal is due to the specific quality of the special organization of Eastern civilizations—a quality which consists in the fact that the public and the moral, i.e., tribal, mode of governance are organically fused together. Unlike in Eastern civilizations, the sharp opposition of statehood to the tribal system in the ancient world was also manifested in the opposition of the ideal to the spiritual as the animistic cults of tribal gods. The animization of the ideal in ancient Greek philosophy, unlike in Eastern philosophy, is opposed to and detached from the animized sensual-visual images of nature in natural philosophy—which is manifested through the "doubling of the world" in the philosophy of Plato.

Upon the discovery of the ideal, the tradition of ancient Greek philosophical thought introduces the idea of law, whereas Eastern thought, in uniting the spiritual and the ideal, animizes their unity as the activity of the spirit—a unity which obeys the law as the logos. E.P. Blavatskaya observes that

> "The 'Manifested Universe'... is pervaded by duality, which is, as it were, the very essence of its ex-istence as 'manifestation.' But just as the opposite poles of subject and object, spirit and matter, are but aspects of the One Unity in which they are synthesized, so, in the manifested Universe, there is 'that' which links spirit to matter, subject to object.
>
> This something, at present unknown to Western speculation, is called by the occultists Fohat. It is the 'bridge' by which the 'Ideas' existing in the 'Divine Thought' are impressed on Cosmic substance as the 'laws of Nature.' Fohat is thus the dynamic energy of Cosmic Ideation [thought-basis]; or, regarded from the other side, it is the intelligent medium, the guiding power of all

> manifestation, the 'Thought Divine' transmitted and made manifest through the Dhyan Chohans, the Architects of the visible World. Thus from Spirit, or Cosmic Ideation [thought-basis], comes our consciousness; from Cosmic Substance the several vehicles in which that consciousness is individualized and attains to self—or reflective—consciousness [the self-consciousness of—reflective—consciousness]; while Fohat, in its various manifestations, is the mysterious link between Mind and Matter, the animating principle electrifying every atom into life."[11]

One can see in the above passage that Fohat denotes an intuitive idea of the natural direction in the development of matter towards the emergence of consciousness. And this direction is personified as an image animized into a reality. The further structuring and content of the "Secret Doctrine" consistently follows from these fundamental attempts and leads to an idea of the hierarchical structure of "Spirit-Matter." Such structuring, which is a special form of speculative knowledge developed by many generations of thinkers—Eastern contemplators—constitutes a pre-scientific and pre-philosophical intuitive-metaphorical system of knowledge. The logic of this system concerns the structuring, and not the substantiation, of its statements, which does not make it [the system] qualify as scientific, while the absence of self-reflexivity does not make it qualify as philosophical. Let us consider in more detail how the content unfolds in the "Secret Doctrine."

The first secret [or legend] posits the cosmic rhythm of Being and Non-Being, in which the Universe is eternal. It is periodic in the incessant appearance and disappearance of worlds and is eternal as a whole.[12]

The second secret responds to the question on the duration of these periods and on the impulse for the re-birth of Cosmos from Non-Being. " 'When all Being dissolves into Non-Being, something indestructible yet continues to exist. This is the Great Cosmic Principle, the Causeless Cause of Non-Being, which after Mahapralaya (the Great Eternity of Non-Being—the author's note) will evoke new manifestations of the Universe, just as after the fire dies out, after it dissolves into Non-Being, 'the principle of fire' will endure"[13] and make possible the reappearance of fire, provoking it out into being. This position corresponds to a kind of synthesis of the divine principle and the law [the logos] in European thought. It is designated "Parabrahman," that which lies beyond Brahman or Cosmos. In religion "Parabrahman" as an idea is personified into the idea of "One God," the Creator of Earth and Heaven. By a similar procedure of animization, the original concept of the One or the Absolute was reduced to "God as Personality," to "The Lord of the Universe," who has a certain character and accepts offerings. However, such a "God" is not known to the ancient legends of the East.

The third secret of the esoteric teaching answers the question of how Cosmos is born after Mahapralaya, whether it arises all by itself, without any outside assistance, or someone creates it. The First Cause of Cosmos, the Great Divine Entity, called Logos, arises first with respect to Being. It is "the word that resounds out of the silence. It is the vibrations or motions of the divine energy, which is at once Light, for Light is the motion of matter. This Light also means the Divine Thought which gives rise to the longest period of the creation of the Universe."[14] The Logos is the "Creator of the Universe" in the sense of an architect who has subordinated a host of builders creating all forms after these ideas of the Logos as the Universal Mind. The ancient legend narrates that after Pralaya the Universe is being built very slowly, over many hundreds of millions of years. On the theosophical teaching, animistic spiritual entities—the laws of creation—guide the evolution of the Universe.

The fourth secret reveals the substance of Cosmic Matter from which [all] worlds are created. The primary "raw material" of cosmic matter is Protocosmic substance—directed matter called Mulaprakriti, i.e., Root-Matter, from which all kinds of matter are created, from the subtlest to the coarsest. The legends distinguish seven states of cosmic Spirit-Matter—similar to steam, water, and ice being states of one kind of matter. Of these seven, only the one known as the lowest and coarsest is visible to the physical eye—the matter of the physical world.

The atoms of the first—the subtlest—state of Spirit-Matter drill holes inside the Protocosmic substance by swirling at high speed. These are the vortices of life enclosed in the thinnest shell—"hollows" filled with the energy of the Logos. The substances of the primary atoms form Spirit-Matter of the highest plane called the "World Divine."

On the next—second—plane, the Logos builds atoms around the atoms of the first plane, forming spiral vortices. The atoms thus created form the second plane called the "Monadic World." The planes of Spiritual Matter that follow arise in a similar manner: the third is called the World of Spirit or Nirvana, the fourth the World of Happiness or the World of Intuition, the fifth the World of Thought or the Fiery World, the sixth the Subtle World (of feelings, emotions, and desires), and the seventh the Dense World. That the idea of Spirit-Matter is animized in this construction seems obvious: "Even the Logos, the Supreme Ruler of life, manifests itself as enclothed in the Universe, which serves him as a form. And the same thing is repeated everywhere, down to the smallest atom."[15] The peculiarity of such a speculative construction of the genesis of the Universe consists in the fact that the ideas of material atoms, their energy and spirit as life force are organically fused together in it. Such views are prerequisites for science, which preserves the ideas of matter and energy, yet introduces the concept of regularity in place of the Spirit. Whereas preserving the animized spiritual as a principle leads to mysticism, occultism, and magic.

The fifth secret reveals the formation of star systems. The theosophical teaching proceeds from the fact that all worlds and planets are individual beings that are the dense houses of the animizing Minds or Planetary Spirits. "Every celestial body is a temple of one of the divine beings, every star is a sacred Abode."[16]

The sixth secret is about life in various Worlds. Life is understood as the energy of the Logos that creates its forms of manifestation from the matter of all the planes. "These forms are created from all sorts of combinations of previously created Cosmic Matter. Innumerable hosts of Beings called Builders, including the so-called Spirits of Nature, take part in their creation."[17] For example, our planet consists of three worlds: the physical part (the Dense World), the subtle part (the World of Feelings, Desires, and Emotions), and the world of thought (the World of Fire). Humanity lives in these three worlds, and when it completes its evolution on Earth, it will move to another planet as superhumanity, and our animal kingdom will begin human life, and our vegetable kingdom animal life.

The seventh secret reveals the history of the origin of planet Earth. The new nebula, from which the Earth was formed, drew in the old planet (Moon) and the latter's water and volatile substances passed in a gaseous state at a high temperature onto planet Earth. The evolution of the planet of Venus is one step in advance of that of the Earth, and its average population is near the adept level. That is why, "the adepts from Venus came to help the inhabitants of planet Earth at its commencement—as Lords, Manus, Buddhas, and other great leaders of evolution."[18]

In the legends the Earth is compared to a large animal that has its own individual life, hence its own mind or its own manifestation of the spirit.

There is one law of birth, growth, and destruction of everything in the Cosmos, from the Sun to the firefly crawling in the grass. There is continuous improvement work with each new manifestation, though Substance-Matter and the Forces remain the same.[19]

As we can see, the esoteric picture of the genesis of the universe contains as prerequisites images that would later form the basis of science, art, religion, and philosophy. And although animism in this esoteric teaching is presented in the classical form, it cannot be separated by using the tools of theosophy from other images animized along with the ideas of spirits. Identification of animism as a special type of thinking occurs when the spirit, being directed at itself, recognizes itself as spirit, and distinguishes itself from other images in which it does not recognize itself as such. These images include Matter and Law or Logos. And whereas theosophical thinking does not separate matter from spirit—does not oppose one to the other as different, mature forms of religion and philosophy do so.

The spirit as a signification attributed to matter as a sign (or as the meaning of a symbol) polarizes one animized image into two components which cannot exist without each other and at the same time lead to the question of their interrelationship—the question of primacy and secondariness. However, firstly, such a raising of the question is possible only with respect to ideal images, provided by reflection on animistic images, and, secondly, ideal images as a product of reflection on animistic ones can be known, are distinct from each other, and can therefore be individually animized as fundamental and primary—as matter (materialism) and spirit (idealism).

However, in theosophical thinking, based on animism, these images coexist as coarse and subtle matter. The gnoseological basis for such coexistence is that animistic thinking, like practico-material thinking, is involved in reality, and, unlike it, aims to capture the whole in addition to the parts, and the essence in addition to the phenomena. Theosophy is the highest form of animistic thinking proper within which "splitting" and differentiation become apparent. The subsequent forms of development of spiritual culture carry within themselves merely transformed forms of animism, which we have called animization. And although religion, art, and science do overcome animism as a mode of human thinking on the basis of the spirituality, conventionality, and literalness of images, respectively, philosophy does so most consistently and effectively through self-reflection.

The uniqueness of the combination of prerequisites for various forms of consciousness in theosophy[20] is revealed in the excerpts provided by E.P. Blavatskaya from private commentaries. They state that the initial Being (after the Mahapralaya that follows every age of Brahma) is

> *a* CONSCIOUS SPIRITUAL QUALITY. *In the Manifested* worlds *(Solar Systems) it is, in its Objective subjectivity, like the film from a Divine Breath to the gaze of the entranced seer. It spreads as it issues from Laya throughout Infinity as a colourless spiritual fluid. It is on the Seventh Plane, and in its Seventh State in our Planetary World.*
>
> *It is Substance to* OUR *spiritual sight. It cannot be called so by men in their Waking State; therefore they have named it in their ignorance* "GOD-SPIRIT."
>
> *It exists everywhere and forms the first Upadhi (Foundation) on which our World (Solar System) is built. Outside the latter it is to be found in its pristine purity only between (the Solar Systems or) the Stars of the Universe, the worlds already formed or forming; those in Laya resting meanwhile in its bosom. As its substance is of a different kind from that known on Earth, the inhabitants of the latter, seeing* THROUGH IT, *believe in their illusion and ignorance that it is empty Space. There is not one finger's breadth (angula) of void Space in the whole Boundless (Universe)....*[21]

It is further revealed that the structure of Matter or Substance is "septenary within our World" and beyond, and each state of matter is graduated into seven degrees of density. The description of these states is interesting in virtue of the possibility to draw analogies with the states of matter as per contemporary cosmology data: from vacuum through the cosmological singularity, the birth of the four physical (fundamental) interactions, and the subsequent atomic-molecular evolution. At the same time, theosophical constructs of material organization include both the conventionality of figurative comparisons characteristic of art and the spiritual qualities peculiar to a religious worldview. In other words, the aforementioned features in theosophy act as prerequisites for science, art, and religion. Identifying these qualities as prerequisites, however—that is, reflecting on these qualities—is the very ground for the emergence of philosophical thinking, the essence of which is, at its origin, in the answer to the question of how thinking—in science, art, and religion, respectively—that is different from animistic thinking is possible.[22] Theosophy, being undivided, non-differentiated according to its qualities, which formed the basis of the subsequent spiritual culture, is perceived by E.P. Blavatskaya as a synthesis of philosophy, science, and religion. However, it is impossible to synthesize that which is undivided and has not therefore attained its developed forms of manifestation. At the same time, a synthesis of those forms of spiritual culture—a synthesis which would reveal the grounds of these forms—would presuppose that these reached such a mature state that their own very grounds, when transformed, proved capable of unification. The leading role of philosophy in this process would not consist in developing within itself and offering some kind of unified, abstract system that could serve as the basis for synthesis, but in revealing the specific conditions under which would be possible various forms of spiritual culture and their specific types of thinking which would constitute the basis for their unity.[23]

Theosophy distinguishes between two worldviews: the world as absolute (unchanging and one in existence) and as relative (changeable and diverse). Viewing the world as relative, as opposed to viewing it as absolute, is an "illusion." Animism is a form of endowing the world with images of mind without recognizing these images as such. However, the traditional understanding of animism, which came from E. Tylor, is associated with the opposition of images of nature to those of spirits. The relation between them develops into the relation between images of matter and images of mind, in the absence of reflection on both as images of thought.

What is difficult about the philosophical analysis of the foundations of the theosophical doctrine is that the properties of the fundamental images take the latter beyond the "limits" of philosophical thinking proper, which essentially rests on the gnoseological aspect of consideration. In the established tradition of philosophical research, if ideal images, the origin of which is revealed by means of gnoseological analysis, are ascribed properties—for example, of value or existence—that go beyond the gnoseological aspect, then they respectively belong to axiological and ontological aspects. However, the ascription in philosophical thinking of the property of existence to an image preserves this image as an element of the thinking subject, whereas in theosophical thinking the image is taken outside the subject. In other words, in philosophy, spirituality (value) and existence are *beyond the gnoseological aspect* (i.e., in the axiological and ontological aspects), whereas in theosophy these are *beyond the limits of thinking itself*. This is an animizing type of thinking leading to the idea of intellectual substantiality, in addition to subjective intellectuality. For this reason, the traditions established in modern philosophy of considering the problem in the ontological, gnoseological, and axiological aspects are not enough to study theosophical teachings.

Analysis of theosophy shows that its content preserves not only rhetorical figures and poetic metaphors, but also animistic images overcome in the European philosophical tradition formed by Plato and Aristotle. At the same time, the study of the mystical teachings of the East contributes to the expansion of the conceptual apparatus and the logico-gnoseological instruments of traditional modern European philosophy.

The emergence of modern European science—an emergence which constitutes the latter as entrepreneurship—democratizes the sphere of thinking, developing publicly accessible and ordered forms of thought. "Without this last feature [development] science has no general intelligibility, and has the appearance of being an esoteric possession of a few individuals—an esoteric possession, because in the first instance it is only the essential principal or notion of science, only its inner nature that is to be found; and a possession of a few individuals, because, at its first appearance, its content is not elaborated and expanded in detail, and thus its existence is turned into something particular. Only what is perfectly determinate in form is at the same time exoteric, comprehensible, and capable of being learned and possessed by everybody."[24]

The above study has found occultism and theosophy to represent a pre-philosophical thinking that preserves the features of animism, yet contains symbolism. It allows us to view this type of thinking as a mediating step between animistic and conceptual, abstract-logical thinking. The principled difference between science as an intra-philosophical phenomenon and occultism, however, is that the former is mediated by philosophy, which develops within itself the logical form of the present existence of thought as the ideal. Plato's dialogues exemplify the process of overcoming animistic ideas in mythology and of the formation of the ideal (concepts).

Notes

1. *Encyclopedic Dictionary of Philosophy*, eds. L.F. Ilyichev, P.N. Fedoseev, S.M. Kovalev, and V.G. Panov (Moscow: Sov. Entsiklopedia, 1983), 680.
2. See S.V. Tukholka, "Occultism and Magic," in *The Heart* 1 (St. Petersburg, 1992), 39–41.
3. Barlet stated that while "positive science" takes its place as "one natural division in all of science," the other division is the so-called occultism. He further classified the structure of the occult (as follows): He designates the first of its three parts as "esotericism" .... The second part is called by him "the theory of magic art" ... and, finally, the third part the application of occult sciences. In order to distinguish the division of occult science from the positive sciences, Barlet proposed the following designations: theognosia—knowledge of the divine world—cosmognosia—knowledge of the cosmic world, and anthropognosia—knowledge of the human world (see François-Charles Barlet, *Occultism* (Leningrad, 1991)).
4. *Encyclopedic Dictionary of Philosophy*, eds. L.F. Ilyichev, P.N. Fedoseev, S.M. Kovalev, and V.G. Panov (Moscow: Sov. Entsiklopedia, 1983), 680–681.
5. E.P. Blavatskaya (1831–1891) "was one of the most extraordinary and controversial figures in the 19th century. The impact of her life, works and teachings on world thought was significant" (*The Occult* [Esoteric] *World of E.P. Blavatskaya* [Madame Blavatsky], compiled by Daniel Caldwell (Moscow, 1996), 11). Currently, there is a

conscious need for a philosophical analysis of theosophy and esotericism (see T.V. Krainev, "Theosophy as Philosophy," in *The Twenty First Century: Russia's Future In a Philosophical Dimension: Proceedings of the Second Russian Philosophical Congress* (June 7–11, 1999), vol. 4, *Philosophy of Spirituality, Education and Religion*, part 2 (Ekaterinburg, 1999), 290–291; V.M. Rozin, "The Esoteric World," *Social Sciences and Modernity* 4 (1992); and G.G. Kopylov, "Esoterics, Science and Extrasensory Perception: How One Can Do Without 'Pseudosciences,'" *Philosophy of Science* 2 (2001)).

6. G.W.F. Hegel, *Lectures on the History of Philosophy*, book 1 (St. Petersburg, 1993), 167. Translator's note: See the English translation of the quotation from G.W.F. Hegel, *Lectures on the History of Philosophy* (1805–6), trans. E.S. Haldane (1892–6), web: https://www.marxists.org/reference/archive/hegel/works/hp/hporiental.htm#indian; with my alternative translation from the Russian in square brackets.
7. Ibid.
8. Ibid., 181.
9. See E.P. Blavatskaya, *The Secret Doctrine: The Synthesis of Science, Religion and Philosophy*, vol. 1, *Cosmogenesis* and vol. 2, *Anthropogenesis* (Leningrad, 1991).
10. E.P. Blavatskaya, *The Secret Doctrine: The Synthesis of Science, Religion and Philosophy*, vol.1, *Cosmogenesis*, part 1, "Cosmic Evolution" (Leningrad, 1991), 51. Translator's note: English translation of the quotation (with my translation in square brackets) from H.P. Blavatsky, *The Secret Doctrine: The Synthesis of Science, Religion, and Philosophy*, vol. 1, Cosmogenesis (Pasadena, California: Theosophical University Press, 2014), 16, web: https://www.theosociety.org/pasadena/sd-pdf/SecretDoctrineVol1_eBook.pdf.
11. Ibid., 50–51. Translator's note: See English ed., 15–16.
12. See *Cosmic Legends of the East*, authored and compiled by S.V. Stulginskis (Moscow, 1992), 9–14.
13. Ibid., 15.
14. Ibid., 18.
15. Ibid., 23.
16. Ibid., 28.
17. Ibid., 29.
18. Ibid., 37.
19. Ibid., 38.
20. See A.E. Nazirov, "Evolutionary Views in Theosophy," in *The Interaction of Science, Philosophy and Theology in the Formation of the Ideas of Evolution in Nature and Society* (St. Petersburg, 1999).
21. See E.P. Blavatskaya, *The Secret Doctrine*, vol. 1, part 1 (Leningrad, 1991), 350–351. Translator's note: See English ed.: 289. Only some words in the quote are fully capitalized to reflect A.E. Nazirov's original.

22. In this regard, the essential feature of philosophical thinking proper is its anti-animistic orientation, which consists in treating images not as animistic, but as ideal. The ideal as a product of reflection on the animistic expands the possibilities of modelling different types of thinking, including the prerequisites for philosophical thinking itself. Philosophy, since it reflects the emergence of various forms of modern spiritual culture, each of which overcomes animism in its own specific manner, originally becomes a kind of a super-cultural phenomenon. This fact has manifested itself in the fates of philosophers, who have come into conflict with culture.
23. Thus, for example, the limitation of philosophy consists in its desire for self-preservation, manifested in its self-containment when trying to construct a philosophy of science and other forms of spiritual culture. In this case, there always arise some prerequisites and conditions that do not fit into philosophy as a type of thinking. This fact received special attention from representatives of both classical (I. Kant) and non-classical (E. Husserl) philosophy (See Edmund Husserl, "The Crisis of European Sciences and Transcendental Phenomenology," *Questions of Philosophy* 7 (1992). On the conditions of the possibility of philosophy, see Karl-Otto Apel, "Transcendental-Hermeneutic Concept of Language," *Questions of Philosophy* 1 (1997).
24. G.W.F. Hegel, *The Phenomenology of Spirit* (St. Petersburg, 1992), 6–7. Translator's note: See English ed.: G.W.F. Hegel, *The Phenomenology of Mind*, trans. J.B. Baillie (Mineola, NY: Dover, 2003), 7.

Chapter Three

The Problem of Prerequisites for a Scientific Theory in the History of Philosophy

Philosophy as reflection on the animistic content of thinking answers the question of being in the form of various fundamental metaphysical positions that determine the mentality of the epoch that pervades the spiritual culture of the time.

According to K. Jaspers, in the "Axial Age," dated from about 800 to 200 years BC, there occurs a breaking through magic and mythology realized by the philosophers of ancient Greece, the Old Testament prophets, and the religious reformers of Iran, India, and China.[1] In ancient Greek philosophy the overcoming of magic and mythology is accomplished by answering the question of being. The cultural transformation in ancient Greece is associated with the process of replacing mytho-poetic ideas by speculative constructs of nature (water, air, apeiron, and fire) that form the basis of natural-philosophical and scientific thinking. The appearance of Platonic eide as the ground of being prepares thought for being formed by the instruments of logic and mathematics in the works of Aristotle and Euclid, respectively. Orientation to "logical consistency, theoretical justification, and the study of cause-and-effect relations and regularities in nature—the most important constituent part of the cultural revolution in ancient Greece—remains the core and pivot of the historical development of the European spirit."[2] The dialogues of Plato contain the prerequisites for both the formal logic of Aristotle and the dialectical logic of Hegel. Aristotelian logic allowed for the axiomatization of mathematics in the form of Euclidean geometry and became the basis for the development of European science.

Medieval philosophy, which projected metaphysics toward the concept of the Christian God—the supernatural and supraintelligent principle—put forward the problem of universals. Conceptualism, which overcame the extremes of nominalism and realism and proceeded from the manifestation of the universal in the particular, became a logical prerequisite for modern European science. Thomas Aquinas' idea of the two truths of faith and reason, i.e., knowledge postulated and knowledge deducible, anticipates the scheme of scientific theory consisting of principles and the fundamental laws based on the latter—laws from which conclusions are derived by deduction. Moreover, Thomas Aquinas presents the relation of the human being to God as the relation of the real to the ideal, that is, as the relation to the limit. This serves as a prerequisite for both the theory of limits in mathematics and the introduction of ideal objects into modern European science.

In the Renaissance period the logical position of Nicholas of Cusa on God as the identity of the minimum and the maximum was the anticipation of differential and integral calculus (the Newton-Leibniz theorem). Involvement in mathematical natural science after the works of Nicholas of Cusa becomes "God's work" in European culture, and inspires N. Copernicus, J. Kepler, G. Galilei, I. Newton, and R. Descartes. It is known that by virtue of thought experiments G. Galilei came up with ideal models that formed the basis of Newtonian mechanics.

In modern times a subject-object relation comes to supersede the medieval position, and medieval scholasticism "turns" into mathematical logic, which is after Boolean algebras used in modern computer technology and computer science.

The basis of modern non-classical thinking in culture is the uncertainty of being, which comes to supersede the classical ideal of clarity and certainty. Such uncertainty, which suspends the subjective-objective opposition, makes up the content of the teachings of A. Schopenhauer, F. Nietzsche, K. Marx, F. Engels, S. Freud, C.G. Jung, E. Fromm, E. Husserl, and M. Heidegger, and is also the initial principle of contemporary physics (quantum mechanics) and mathematics (the theory of fuzzy sets).

Let us consider the reflection on the foundations of science within the history of philosophy. According to Aristotle, scientific knowledge differs from perception in that its subject matter is the universal and necessary, and not the isolated and accidental, and from opinion in that reasoning is always true.[3] Ancient philosophers, in comprehending elements of scientific—in particular, mathematical—knowledge, aimed to come up with such rules of reasoning as would yield reliable knowledge. This tendency found its expression in Aristotle's syllogistic method of deductive reasoning. Such a method, which was used to analyze mathematical proofs and to systematize and classify empirical material, served as a kind of addition to the descriptive method. Unlike dialectics as a special method of cognition (identification and overcoming of contradictions through discussion as understood by Socrates and Plato), these syllogistic and descriptive methods are more concrete in character. Modern authors distinguish two ways in the development of science: the "Babylonian"—describing a large number of concrete facts and establishing interrelations among them—and the "Greek"—an axiomatic identification of deep-seated connections.[4] It can be posited that ancient philosophy already contains some elements of the methodology of scientific cognition in the form of positing the descriptive and syllogistic methods as complementary.

The ideal form of universal-conceptual ancient thinking acquires in patristics its corresponding ideal content through the concept of God as absolute otherness than our world. The cognition of God by means of logic makes up the content of scholasticism. In the Renaissance period, as a result of treating God as not other—that is, as an ideal model of nature—its study employs more precise and rigorous instruments than those of logic—the instruments of mathematics.

F. Bacon shows the attractiveness of science, making it value-oriented and harmonizing "its relations with existing social institutions."[5] The image of science is now to correlate with the requirements of morality, religion, and politics. The project in F. Bacon's *New Atlantis* to institutionalize science anticipated the appearance of European academies of sciences.[6]

The era of the emergence of empirical natural science more deeply and essentially poses the problem of choosing a true method of science. It is known that F. Bacon, identifying three possible ways of knowing—"the way of the spider," "the way of the ant," and "the way of the bee"—chooses the third as the only true one. Noting that the ascent from sensations to general axioms, from empiricism to theory, happens "continuously and gradually,"[7] Bacon thereby points to the need for a synthesis of the empirical and the theoretical in scientific cognition. However, Bacon's epistemological motive consists in the inductivism of this synthesis: without denying the role of rational methods, he considers the data of experience to be the only initial ground for scientific theory, from which laws and scientific principles are derived via inductive generalization.[8]

The opposite way is also possible: from the clear, most general axioms to the data of experience via deductive inference (R. Descartes). Descartes also uniquely synthesizes the empirical and the theoretical method. "Bacon opposed Descartes not so much as an empiricist did a rationalist…but as an inductivist did a deductivist."[9] Using the deductive method, Descartes dreamt of realizing the idea of "pantometry" (universal measurement) by reducing all physics to geometry and geometry to algebra."[10] In order to do so, it was necessary to be guided by some "general" truths, and Descartes takes for true only that the reliability of which is equal to the reliability of his own existence, his own thinking.[11]

The works of Bacon and Descartes shaped the contemplative subject-object model of studying the origin of the theoretical knowledge of science. In the subsequent philosophical writings of the classical period this initially proposed model was merely modified and remained contemplative.

The next form of empirical and theoretical synthesis was developed in the works of T. Hobbes, whose approach can be explained by his aim to combine the opposed requirements of Cartesian and Baconian methodologies. Hobbes considers the methods of both induction and deduction, classifying sciences according to the predominance of one method over the other. Identifying analysis with induction and synthesis with deduction, Hobbes believes that induction prevails in the formation of scientific theory. However, the construction of science, in his opinion, should conclude with the deduction of "complex bodies and their properties." He calls this deduction and synthesis the "*a priori* method."[12] The specific feature of Hobbes' methodology consists in "reversing the method" (from induction to deduction and from analysis to synthesis), and the research path is interpreted as a movement from effects to causes and from causes to more diverse effects. Similarly, contemporary methodology views the axiomatic method as the reverse of the genetic one.

The rationalistic tendency in philosophy, the axiomatic forms of Euclidean geometry and Newtonian physics, was reflected in the culmination of B. Spinoza's teaching. The Dutch thinker regarded the attributes of two Cartesian substances as the two attributes of one substance. This substance is nature (or God as the perfect being in nature), which is according to Spinoza "the cause of itself,"[13] or *causa sui*. The attribute of thinking is ascribed an "absolutely infinite intellect" combining [all] possible acts of mental activity.

Spinoza resolves at the abstract-logical level the question of the origin of theory as the question of the relationship between the attributive properties of substance and modes. From the concept of substance is derived the logical rationale for a specific theory, and not the theory itself, which reflects within itself specific objects. He notes that "to perceive them all at once would by far exceed the powers of the human intellect."[14]

However, Spinoza points out the nature of the formation of the "world of modes:" the finite possibilities are realized in the bosom of the infinite through the latter's limitation (logical determination).[15] In his discourse on modes, Spinoza relies not only on the concept of substance, but also on the data of empirical experience. (So does Hegel, only much later.) Spinoza has an implicit conclusion that "to perceive" all single things in theories would require an endless process of the development of empirical and theoretical cognition, involving many generations of people.[16] As a metaphysician, Spinoza believed that clear [and distinct] cannot follow from obscure [and confused] perception (sensory experience), absolute from relative, and thus detached empirical data from theoretical studies (rational intuition).

The methodology of seventeenth-century rationalism (the acceptance of only absolute, necessary, universal, and universally binding truths) also guided G. Leibniz in formulating his methods of cognition. He believed that the truths underlying a theory are derived from the mind and that the laws of the world are reduced to the laws of formal logic. Therefore, Leibniz's methodological principles are at the same time ontological.

Leibniz formulates new principles for building his own philosophy—the principle of universal discernibles and the identity of indiscernibles. He tends to identify real causes with the logical principles of theory. Allowing that there is middle (possible) knowledge between the false and the true, Leibniz observes that reality is opposed to possibility as existence is opposed to essence.[17] Hence, according to Leibniz, the world as a whole is "accidental" and other differently arranged physical worlds are possible. He universalized the laws of Aristotle's formal logic and introduced the principle of sufficient reason, believing that the truth or falsity of any statement or theory depends only on a certain principle. Here arises the problem of the primacy and secondariness of principles, or their hierarchy. According to Leibniz, the first principles of truths are rooted in "one universal mind" (philosophical truths) and the second principles in individual monads (concrete scientific truths). Thus, the relation of philosophical knowledge and the principles of scientific theory is that of the highest monad and individual monads.

I. Kant was the first to perform a special analysis of the principles of scientific—in particular, physical—theory. Combining the empirical and theoretical methods in a peculiar manner, he showed that any physical theory has elements—"transcendental presuppositions"[18]—that cannot be derived from experience. Transcendental or pre-experimental presuppositions make experiment possible. The science that studies the origin and meaning of knowledge, Kant calls transcendental logic, which is divided into analytics and dialectics. Kant criticizes Leibniz for the fact that he, having no transcendental topic at his disposal, allows himself to be "deceived by the amphiboly of the concepts of reflection [in] construct[ing] an intellectual system of the world… by comparing all objects only with the understanding and the abstract formal concepts of its thinking."[19] In Kant's own language, the relation of theory and empiricism is that of "pure apperception" and "empirical apperception." The origin of "pure apperception," however, is explained from "transcendental apperception" (a non-empirical principle).[20] According to Kant, the nature of the fundamental principles of physical theory is derived from *a priori* knowledge, similar to the "innate ideas" of Descartes and Leibniz. However, Kant's merit is that he is the first to have made a clear distinction between the empirical and the theoretical and between "understanding" and "reason" and offered a solution to the problem of the fundamental principles of theory by appealing to the concept of *a priori* knowledge.[21]

Using the concepts of understanding and reason, Hegel was the first to have divided the methods of cognition between philosophy and particular sciences. He argued that the previous metaphysical systems employed for philosophical cognition the methods of particular sciences, for example, the analytical and synthetic methods. "That these methods, however indispensable and brilliantly successful in their own province, are unserviceable for philosophical cognition is self-evident. They have presuppositions; and their style of cognition is that of understanding, proceeding under the canon of formal identity."[22] From this standpoint Hegel criticizes Spinoza, Wolff, and Kant. In his view, any particular science, using its own method, "in its course," necessarily goes beyond its own limits—"beyond the principle of the understanding." Further movement occurs due to the philosophical method, which contains the analytical and the synthetic method merged together and consists in the development of the Idea. Then the movement of cognition returns to its "original province" (particular sciences) and complements it (them) with new concepts. A concept is formed as an other of the Idea, as a recognition of the latter's totality.[23] Similarly the principles of scientific theories arise as some general concepts reflected upon within the development of the Idea.

We can posit that according to Hegel, philosophical ideas are the principles of scientific theory. However, his view that these ideas form all the concepts of theory (without the participation of concrete-scientific concepts themselves) is natural philosophical in character. At the same time, Hegel, unlike his predecessors, combines the empirical and the theoretical method of cognition in a more dialectical manner. He advances the idea that any principle of theory is an undeveloped result, while the result is a developed principle.

Positivism made a kind of protest against both the speculative nature of natural philosophy and the absolutization of extreme empiricism (in its Humean sense). Rejecting the specific nature of philosophical knowledge, positivism identified the latter's methodology with the methodology of the special sciences, viewing philosophy as the summation of natural scientific knowledge in its most general form. Positivist criticisms of speculative metaphysics turned into criticisms of philosophy itself and of the philosophical-ideological comprehension of both scientific knowledge and scientific methodology.

The approach of positivism to cognition is tied to its goal to construct a metatheory, which reveals itself as possessing the features of the natural-philosophical "science of sciences." Examples of the schematism of positivist thinking are A. Comte's "law of the three stages of development,"[24] the conclusion revived by H. Spencer that scientific theories require original, pre-experimental, fundamentally unknowable "principles," and J.S. Mill's system of inductive logic.[25]

E. Mach subjected to analysis the very concept of experience (empirio-criticism) and put forward the principle of the economy of thought, which according to Mach has a pre-experimental essence to it. He believed that according to the principle of the economy of thought the principles of theory constitute an economical description of experience data. Here empiricism and Kantian apriorism are uniquely combined in interpreting the origin of the principles of scientific theory.[26]

At the beginning of the twentieth century D. Hilbert proposed a program to develop a mathematical theory of proofs. In the process of its implementation there have emerged many problems involving analysis of the construction of the language of mathematics, the construction of the language of scientific knowledge, the methods of constructing the language frameworks of science, etc. The latter circumstance significantly affected the form of positivism and there arose logical positivism. The principal task became the axiomatization of mathematics (B.

Russell). L. Wittgenstein, in trying to axiomatize all science, associates the principles of theory to "atomic sentences." Representatives of the "Vienna Circle," in proposing the program of the empirical substantiation of theoretical knowledge, put forward the verifiability principle in the first place.

Unlike positivism, neopositivism does not completely reject the significance of philosophy, but tries to divide unitary philosophical knowledge by absolutizing particular logico-semantic tasks and the significance of logic. Neopositivism is mostly interested in the problem of the meaning of linguistic expressions as the ground of the states of consciousness that alone are accessible to cognition. Neopositivism, therefore, considers philosophical activity as an analytical activity in search of the final data of natural (ordinary) or artificial language. Two major forms of neopositivism can be identified from its general evolution: logical positivism and linguistic positivism.

Taking the position of empiricism, logical positivism comes to the rejection of any ontological presuppositions by which are meant all the epistemological and ontological ("metaphysical") problems of philosophy, considering them to be of no cognitive value and meaningless. Of the whole of science, only empirical utterances are singled out and clothed in the form of formal-analytical logico-mathematical propositions.[27]

For the purposes of eliminating meaningless metaphysical utterances and selecting meaningful scientific propositions logical positivism proposes the principle of verification by which to make an empirical detection of the fact that determines the meaning of the utterance. As a result, logical positivism completely reduces the entire content of scientific theories to statements of empirical fact. Thus, not only "metaphysical" problematics, but also all the theoretical propositions of both philosophical and natural science knowledge are rejected.

In line with the ideas of logical positivism various ideas were expressed about the nature of the principles of scientific theory. B. Russell came to the conclusion that the principles of theory are such general propositions as go beyond the limits of experience and are grounded in faith, which is "primitive… implicit in animal behaviour."[28] According to H. Reichenbach, new concepts, principles, and hypotheses are derived by speculation, by trial and error; he claimed that "[t]he method of trial and error is the only existing instrument of prediction [cognition]."[29] Ph. Frank confined himself to stating that the principles of science are a product of human creative ability, a product of human imagination.[30] The truth of a theory, in his judgement, can only be ascertained empirically, after the latter's construal. H. Poincaré emphasizes in cognition the leading role of intuition as a special extra-experimental and super-logical ability of the genius mind without which a new theory cannot be constructed.[31] Similarly, Ph. Frank points to the "extra-experimental" (pragmatic) component included in scientific theory.[32]

Pragmatists drew attention to the constructive nature of the methods of science and their dependence on cognitive tasks.[33] They revealed the connection of research procedures with human activities and specific situations in which the process of cognition is carried out; however, this led them to identify logic with the methodology of research (J. Dewey). One of the contemporary branches of pragmatism—operationalism—treats method as a specific measuring procedure. Its concept is based on the identification of the meaning of a scientific concept with the measurement operations carried out by the researcher.[34]

In neopositivism, for example, the question of the reality of objects is reduced to that of the inclusion of these objects in some system—a "linguistic framework" (R. Carnap). Such an interpretation of reality R. Carnap calls empirical and genuinely "scientific."[35] To substantiate this logical positivism puts forward two propositions: first, the very problem of the reality of objects is pseudoscientific; second, the true object of study is the system of the language of science. Since the language employed in natural science did not fully satisfy the criteria of scientific knowledge, it became necessary to determine such language forms as would satisfy those criteria. As the model for such forms logical positivism adopts symbolic systems from mathematics and logic.

Axiomatic systems, which are the syntactic interpretations of natural science theories, are hypothetico-deductive systems. Such systems were widely used in logical positivism to study the formal-logical structure of theory. However, as is plain to see, on such an analysis, unlike structural elements, such systemic elements of theory as concepts endowed with content by empirical interpretation elude attention.

As was shown by logical positivists themselves, owing to the hierarchy of the elements of the system of theory it is impossible fully adequately to reflect the content of theory by means of empirical interpretation. Hence the inference follows that there is a kind of excess content in theory than is warranted by the empirical basis. In this case, according to the general methodological position of logical positivism, it is possible only unilaterally to study the properties of theory; however, in principle, it is impossible to explain the nature of the "excess" content of theory, and therefore, the problem of the principles of theory turns out to be fundamentally insoluble.

The neopositivist understanding of the principles of theory is associated with the Machist conception of cognition, which is the resulting development of the views not only of Mach, but also of Duhem, Pearson, Ostwald, and a number of other positivists of the preceding epoch.[36] At that stage in the development of positivism when its representatives believed that only sensations formed the ground of cognition, positivism was unable to explain what was behind those sensations. If we accept that sensations arise when affected by something other which lies outside of the sensations (by what could be regarded as the external world), then this statement, or position, can only be verified by comparing some sensations with others. The latter, however, in turn, must be compared with a third set of sensations and so on without end. There arises a paradox according to which within the framework of such gnoseology the nature of sensations remains inexplicable.

Within the framework of logical positivism another general contradiction is revealed. Sensations derived from experience are to take the form of certain utterances about the data of observation. Here arises the question of the relation between the content of sensations and the form of its utterance. In order to verify whether the content of a sensation changes due to [the change of] the form of its utterance, it is necessary to compare this content with [the content of] its utterance in another form. And further the content of the sensation expressed in the second form must be compared with the content of the sensation expressed in a third form, etc.

Thus arises another paradox, the attempts to solve which are characteristic of the third stage in the development of positivism. Schlick assumes that there is only one essence which does not admit division into essence and phenomena; however, this leads to the identification of the content of the sensation with the form of its utterance. In this was manifested the influence of Wittgenstein on Schlick. Wittgenstein declared the world a set of facts, and not things—to wit, facts as presented in logical "space." The logical image of facts is thought, and the thought is expressed through a sign identified with a sentence. Thus, in Wittgenstein there occurs the substitution of reality with a certain logical model.[37] Between this model, taken as the original, and logical models a homeomorphic relation can be established. Thus, according to Wittgenstein, as well as actually Schlick, the logical model is a direct mirror reflection of the original model.

Having at their disposal elementary utterances about sensations, observation data, or facts, the researcher questions the truth of these utterances. In order to verify the truth of individual sentences, one must compare them with other sentences, and these others with a third set of sentences, etc. Thus arises another paradox of logical positivism.

Evidently, in order to solve such a paradox in the context of the gnoseology of positivism, it is necessary to assume the existence of special sentences to which all other sentences could be reduced. Elementary sentences of this kind, Wittgenstein called "atomic," Schlick "sentences of observation,"[38] and Carnap /L0./ language which contains the language of observation and logical sentences that do not contain the terms of theory.[39] Then the fundamental language of science is a more complex formation of simple terms and belongs in the language of theory. The latter circumstance served as the basis for a stricter division of scientific knowledge into empiricism and theory (Ph. Frank).

The methodology of logical positivism accepts that the language of observation consists of three kinds of terms: descriptive—terms that denote the observable properties of objects, terms that denote certain relations, and terms that denote objects and phenomena themselves. R. Carnap tends to believe that the language of physics, with the exception of its mathematical part, remains a natural language. Therefore, it is necessary to find more perfect language forms, similar to the coordinate method of the description of physical processes. However, for the expression of many physical phenomena spatio-temporal description is insufficient, and representatives of logical positivism, such as N. Goodman, well realized that themselves. It turns out that what is required for the analysis of particular concepts of science is the study of their specific ontological status. But it is known that neopositivism rejects the ontological analysis of objects as a meaningless metaphysical problematics.

Thus, at the mature stage of its development logical positivism aims to get rid of both natural and philosophical questions at the same time. The criteria of the scientificity and truthfulness of knowledge proposed by neopositivism are only of an intra-scientific character. These criteria are such as the empirical verifiability of theory, i.e., the reduction of theoretical to empirical terms (Ramsey's theorem), or Carnap's principle of observability, i.e., the reduction of unobservables to observables. The authors of such a program were inspired by the fact that theoretical conceptions of science always allowed for the possibility of translating, to a certain extent, some concepts into others, as was the case, for example, with the field of mathematics when translating the concepts of number theory into those of set theory. However, the general scheme of such translation when applied to all the sciences turns out to be immeasurably more complex.[40]

Ultimately, at this stage of the development of logical positivism—at its final, concluding stage—a situation has arisen according to which a theory is believed to be true and scientific only if its content can be reduced to a certain set of empirical terms, while at the same time only that system of empirical terms is believed to be scientific and true which itself can be assimilated by a theory. This forms a logical circle from which there is no escape. Thus arises another paradox which can possibly only be solved by going beyond the initial gnoseological position of positivism. In this connection, it should be observed that the two ways that have arisen of solving this paradox no longer belong in logical positivism itself.

The author of the first approach was K. Popper.[41] Trying to develop a non-contradictory system within the framework of neopositivism,

Popper comes to the conclusion about its logical inconsistency, which led this scientific methodologist to go beyond the initial propositions of positivism. According to Popper, a theory should be regarded scientific based on the principle not of confirmation but of refutation by empirical facts. In this case, the disproving factors are supplied by another theory which competes with the first. There arises a situation of competing theories.[42] The researcher is faced not with the problem of confirming a theory, but with the problem of choosing between competing theories. The decisive criterion for the scientificity of a theory becomes the possibility of principled refutation (falsifiability). Such theories are considered on the basis of preference. Popper's conception is already more tolerant of philosophical assumptions since the assessment of the preference for theories is associated precisely with such assumptions. The criterion of falsifiability has its own difficulties which limit its scope of application. In particular, for the theoretical propositions of existence the "symmetry" of confirmability and refutability is violated, and hence, the tenability of such propositions cannot be assessed from the viewpoint of falsificationism.[43]

The second approach to solving the last paradox of neopositivism is associated with the explicit recognition of the need for both ordinary and philosophical language in scientific knowledge. Although this approach is represented by a number of authors who are not uniform in the contents of their works, there is something in common that unites them; it consists in the assumption of some presuppositions that anticipate solutions to the problems that arise in science. Therefore, scientific knowledge is viewed as "immersed" in a system of preliminary ideas that guide the researcher. E.A. Mamchur identifies two approaches into which, in her opinion, the new tendency breaks down in the "philosophy of science."[44] The first is represented by the works of T. Kuhn, P. Feyerabend, and N. Hanson, and the second by the works of I. Lakatos.

T. Kuhn developed a peculiar historico-sociological conception of the development of science. He views its functioning as the result of the cumulative intellectual efforts of the "community of scientists." Consequently the results of experience and theoretical studies cannot be disengaged from the cumulative socio-psychological norms of the scientific community and of the scientific era as a whole. Accordingly, the fundamental principles of theory are contained in an "implicit body of intertwined theoretical and methodological belief" that constitutes a "paradigm"[45] as the basis for solving problems of science at the stage of "normal science." The change of scientific theories is due to scientific revolutions which, according to Kuhn, are based on a shift of scientific paradigms incompatible among themselves. P. Feyerabend also points to the need for a radical replacement of the ontology of one theory by the ontology of another in the course of the change and development of theories.[46]

I. Lakatos, unlike K. Popper, regards as a "unit" of scientific knowledge not a particular theory, but a scientific research program covering a set of theories.[47] A scientific research program performs a meta-theoretical function in relation to a set of theories.

The fundamentally new elements of the above alternatives are: (1) expansion of the object of study; and (2) going beyond strictly scientific knowledge, toward broad gnoseological formations, such as a "paradigm," an "intellectual background," a "research program," etc.

However, in its consistent presentation, the post-positivist reconstruction of science encounters its own difficulties, which prove insoluble within the confines of its program. Such difficulties are primarily associated with the identification of the substance and role of the above gnoseological formations and with the problem of their continuity in the development of science. In order to solve these difficulties, it is necessary to properly evaluate and understand the role of social factors in the genesis of science.

Notes

1. See Karl Jaspers, "The Origin and Goal of History," in Karl Jaspers, *The Meaning and Goal of History* (Moscow, 1991), 32–98.
2. B.V. Markov, "Science as the Knowledge and Form of Life," in *The Logic and Development of Scientific Knowledge* (St. Petersburg, 1992), 6.
3. See Aristotle, *The Soul* (Moscow, 1937), xvi–xvii. For more detail on the relation of the teachings of Aristotle to the science of modern times, see K.A. Sergeev and Ya.A. Silin, *The Dialectic of the Categorical Forms of Cognition* (*the Cosmos of Aristotle and the Science of Modern Times*) (Leningrad, 1987).

4. See Yu.I. Kulakov, "The Structure and Unitary Physical Picture of the World," *Questions of Philosophy* 2 (1975): 15 and Richard Feynman, *The Character of Physical Law* (Moscow, 1988).
5. A.S. Kravets, "Socio-Cultural Niche of Science," in *Physics within the System of Culture* (Moscow, 1996), 16.
6. In the nineteenth century there emerge ideas of the intrinsic value of pure science in its close connection with social progress. Under the conditions of developed industrial production science becomes a direct productive force, and in the second half of the twentieth century the scientific community becomes aware of its responsibility for the survival of humanity and the justification of humanistic values (see E.A. Mamchur, "Physics and Etiquette," ibid., 101–103).
7. See Francis Bacon, "The New Organon," in Op. 2 vols., vol. 2 (Moscow, 1972), 14–16.
8. Under the influence of Bacon's methodology, principles as knowledge derived from experiments were opposed to hypotheses. See Isaac Newton, *Mathematical Principles of Natural Philosophy* (Petrograd, 1916), 591.
9. I.S. Narsky, *Western European Philosophy of the Seventeenth Century* (Moscow, 1974), 36.
10. Descartes peculiarly anticipated the tendency of geometrizing physics, charted in the works of Lagrange, Einstein, Wheeler, etc.
11. See the works of René Descartes, "Discourse on the Method," in Op. 2 vols. (Moscow, 1989), 269.
12. See Thomas Hobbes, "Elements of Philosophy. Part One. On the Body," in *Selected Works*, in 2 vols., vol. 1 (Moscow, 1964), 116.
13. See Baruch Spinoza, "The Ethics. Part One," in *Selected Works*, in 2 vols., vol. 1 (Moscow, 1957).
14. Ibid., 353.
15. See I.S. Narsky, *Western European Philosophy of the Seventeenth Century* (Moscow, 1974), 214, 223–224.
16. The creation of contemporary physics showed particular interest in Spinoza (see B.G. Kuznetsov, "Spinoza and Einstein," in *The Einstein Collection* (Moscow, 1968), 28).
17. See Gottfried Wilhelm Leibniz, *New Essays on Human Understanding* (Moscow-Leningrad, 1963), 256.
18. Immanuel Kant, *Prolegomena. Op.*, vol. 4, part 1 (Moscow, 1965), 101. On Kant's role in solving the problem of the foundations of science, see M.S. Kozlova, "The Problem of the Foundations of Science," in *The Nature of Scientific Cognition* (Minsk, 1979).
19. Immanuel Kant, *Critique of Pure Reason. Op.*, vol. 3 (Moscow, 1964), 320. Translator's note: See English ed.: Immanuel Kant, *Critique of Pure Reason*, trans. and ed. Paul Guyer and Allen W. Wood (New York: Cambridge University Press, 1998), 371–372.
20. Ibid., 191, 142.

21. The major followers of Kant were H. Cohen and W. Windelband. In our time Kant's ideas are peculiarly being revived (see Rudolf Carnap, *Philosophical Foundations of Physics* (Moscow, 1971), 47 and N.P. Konopleva and G.A. Sokolik, "Symmetries and Types of Physical Theories," *Questions of Philosophy* 1 (1972): 118).
22. G.W.F. Hegel, *Encyclopedia of the Philosophical Sciences*, vol. 1 (Moscow, 1974), 415. Translator's note: Translation of the quotation from the English edition: G.W.F. Hegel, *Encyclopedia of the Philosophical Sciences*, part 1, p. 146, web: http://hegel.net/en/pdf/Hegel-Enc-1.pdf.
23. Ibid., 423.
24. Auguste Comte, *The Course of Positive Philosophy*, vol. 1 (St. Petersburg, 1903), 3–4.
25. See John Stuart Mill, *A System of Logic, Ratiocinative and Inductive* (Moscow, 1914).
26. See Ernst Mach, *Analysis of Sensations and the Relation of the Physical to the Psychical* (Moscow, 1907).
27. See M.S. Kozlova, *Philosophy and Language* (Moscow, 1972), 18.
28. Bertrand Russell, *Human Knowledge* [*Human Knowledge: Its Scope and Limits*] (Moscow, 1957), 465–466.
29. Hans Reichenbach, *The Rise of Scientific Philosophy* (Berkeley/Los Angeles: California University Press, 1951), 241, 247–249.
30. Philipp Frank, *Modern Science and its Philosophy* (Leningrad, 1950), 276–278.
31. See Henri Poincaré, *Science and Method* (St. Petersburg, 1910), 37–39.
32. See Philipp Frank, *Philosophy of Science* (Moscow, 1960), 500–510.
33. See V.V. Bykov, *Methods of Science* (Moscow, 1974), 19.
34. See Percy Williams Bridgman, *The Nature of Physical Theory* (Princeton, NJ: Princeton University Press, 1963).
35. See Rudolf Carnap, *Meaning and Necessity* (Moscow, 1959), 64–66.
36. See V.A. Shtoff, *Modelling and Philosophy* (Leningrad, 1966), 52–54.
37. See Ludwig Wittgenstein, *Logico-Philosophical Treatise* (Moscow, 1958); M.S. Kozlova, "Logic and Reality," *Questions of Philosophy* 9 (1965) and *Philosophy and Language* (Moscow, 1972); and A.F. Gryaznov, *The Evolution of L. Wittgenstein's Philosophical Views* (Moscow, 1985).
38. Moritz Schlick, "The Foundation of Knowledge," in *Analytic Philosophy. Selected Texts* (Moscow, 1993), 45.
39. See Rudolf Carnap, *The Philosophical Foundations of Physics* (Moscow, 1971), 341.
40. See V.V. Tselischev, *Logical Truth and Empiricism* (Novosibirsk, 1974), 15. On the criteria of scientificity, see B.V. Ilyin, *Criteria of the Scientificity of Knowledge* (Moscow, 1989).
41. See Karl Popper, *Logic and the Growth of Scientific Knowledge* (Moscow, 2004). Transltor's note: See English ed.: Karl Popper, *Conjectures and Refutations: The Growth of Scientific Knowledge* (London/New York: Routledge, 2002).
42. See E.A. Mamchur, *The Problem of Choosing a Theory* (Moscow, 1975), 104–117.

43. See V.S. Gryaznov, "On Popper's Criterion of Scientific Knowledge," in *Problems of the Logic, Methodology and Philosophy of Science* (Moscow, 1972).
44. E.A. Mamchur, *The Problem of Choosing a Theory* (Moscow, 1975), 169. For an analysis of the various tendencies, see *Critique of the Contemporary Non-Marxist Conceptions of the Philosophy of Science*, ed. A.I. Rakitov (Moscow, 1987).
45. Thomas Samuel Kuhn, *The Structure of Scientific Revolutions*, 2nd ed. (Chicago and London: University of Chicago Press, 1968), 13–17. Translator's note: Quote from pp. 16–17, web: https://projektintegracija.pravo.hr/_download/repository/Kuhn_Structure_of_Scientific_Revolutions.pdf.
46. Paul Karl Feyerabend, "Explanation, Reduction, and Empiricism," in *Minnesota Studies in the Philosophy of Science*, vol. 3, eds. Herbert Feigl and Grover Maxwell (Minneapolis: University of Minnesota Press, 1962); Paul Karl Feyerabend, *Selected Works in the Methodology of Science* (Moscow, 1986).
47. See Imre Lakatos, *Proofs and Refutations* (Moscow, 1967); *Falsification and the Methodology of Scientific Research Programs* (Moscow, 1995); and "The History of Science and its Rational Reconstructions," in Thomas Samuel Kuhn, *The Structure of Scientific Revolutions* (Moscow, 2003).

Conclusion

Notwithstanding that neopositivists have widely used logical and mathematical methods of analysis of theory and have obtained certain results in the study of its logical structure, as well as in the syntactic and empirical interpretations of theory, their subjective-objective gnoseological model of scientific knowledge constrains further research progress. In the late works of K. Popper, as well as in the works of contemporary postpositivists, one notes an expansion of the gnoseological model due to the inclusion of certain social factors in the object of study; however, most consistently, and much earlier, these problems were raised and substantiated in the Marxist methodology of scientific cognition, which overcame the contemplative nature of preceding natural-philosophical conceptions by introducing in the structure of subjective-objective relations the concept of material-objective human activity (Praxis). The introduction of the concept of Praxis as a mediating relation between the concepts of the ideal and the natural-material marked the transition from classical to non-classical philosophy.

Bibliography

English Sources

Bridgman, Percy W. *The Nature of Physical Theory*. Princeton, NJ: Princeton University Press, 1963.

Brinton, Daniel G. *Religions of Primitive Peoples*. New York/London: G.P. Putnam's Sons, 1897.

Eliade, Mircea. [*The Myth of the Eternal Return: Or,*] *Cosmos and History*. Moscow: Progress, 1987.

——— *Shamanism: Archaic Techniques of Ecstasy*. Princeton, NJ: Princeton University Press, 1972.

Feyerabend, Paul K. "Explanation, Reduction, and Empiricism." In *Minnesota Studies in the Philosophy of Science*, vol. 3, edited by HerbertFeigl and GroverMaxwell, 28–97. Minneapolis: University of Minnesota Press, 1962.

Frank, Philipp. *Modern Science and its Philosophy*. Leningrad, 1950.

Kuhn, Thomas S. *The Structure of Scientific Revolutions*. 2nd ed. Chicago/London: University of Chicago Press, 1968.

Reichenbach, Hans. *The Rise of Scientific Philosophy*. Berkeley/Los Angeles: California University Press, 1951.

Russian Sources

Авеста. Сост. Рак И.В. СПб., 1997.

Автономова Н.С. Рассудок, разум, рациональность. М., 1988.

Апель К.-О. Трансцендентально-герменевтическое понятие языка // Вопр. философии, 1997, № 1.

Аристотель. О душе. М., 1937.

Ахутин А.В. Открытие сознания (древнегреческая трагедия) // Человек и культура. М., 1990.

Барле Ф.-Ч. Оккультизм. Л., 1991.

Бауэр Э.С. Теоретическая биология. М.-Л., 1935.

Башляр Г. Новый рационализм. М., 1987.

Блаватская Е.П. Тайная доктрина.: Синтез науки, религии, философии. Т.1. Ч.1. К осмическая эволюция. Космогенезис. Л., 1991. Т. 2. Антропогенезис. Л., 1991.

Быков В.В. Методы науки. М., 1974.

Быстрицкий Е.К., Кошарный С.А. Герменевтика: гносеологические и онтологические аспекты критического анализа // Филос. науки. 1987. № 1.

Бэкон Ф. Новый Органон // Соч. в 2-х т. Т. 2. М., 1972.

Василенко Л.И. Магия: старое зло или новое благо? // Вопр. философии. М., 1994. № 2.

Витгенштейн Л. Логико-философский трактат. М., 1958.

Владимиров Ю.С. Фундаментальная физика, философия и религия. Междунар. инновац. исслед. центр альтернатив. образоват. и соц. технологий. Кострома, 1996.

Выготский Л.С. Проблемы психического развития ребенка // Избр. психологические исследования. М., 1966.

Гадамер Х.-Г. Истина и метод. Основы философской герменевтики. М., 1988. Герменевтика: история и современность. М., 1985.

Гайденко П. П. Научная рациональность и философский разум. М., 2008. От исторической герменевтики к герменевтике бытия. Критический анализ эволюции М. Хайдеггера // Вопр. философии. 1987. № 10.

Гегель Г.В.Ф. Лекции по истории философии. Книга 1. СПб., 1993.

——— Феноменология духа. СПб., 1992.

——— Энциклопедия философских наук. Т.1. М., 1974.

——— Философия духа. Т.3. М., 1977.

Гоббс Т. Основы философии. Часть первая. О теле // Избр. произв. в 2-х т., Т. 1. М., 1964.

Гончаренко Н.Е. Гений в искусстве и науке. М., 1991.

Грязнов А.Ф. Эволюция философских взглядов Л. Витгенштейна. М., 1985.

Грязнов Б.С. О попперовском критерии научного знания // Проблемы логики, мет одологии и философии науки. М., 1972.

Гумилев Л.Н. Этногенез и биосфера Земли. Л., 1990.

Гуревич П.С. Видный мыслитель XX столетия // Фромм Э. Душа человека. М., 1992.

Гуссерль Э. Кризис европейских наук и трансцендентальная феноменология // Вопр. философии, 1992, № 7.

Даль В.И. Толковый словарь живого великорусского языка. Т. 2. М., 1989.

Деррида Ж. Шпоры: Стили Ницше // Филос. науки. М., 1991. № 2, 3.

Елизаренкова Т.Я. Ригведа. Мандалы I–IV. М., 1989.

——— Словарь основных мифологических персонажей и ритуальных понятий // Ригведа. Мандалы I–IV. М., 1989.

Ильенков Э.В. Философия и культура. М., 1991.

Ильин В.В. Критерии научности знания. М., 1989.

Ильин В.В., Мариничев Э.А. Гносеологические аспекты проблемы понимания в естественнонаучном и гуманитарном знании // Естественнонаучное и социогуманитарное знание: Методологические аспекты взаимодействия. Л., 1990.

Исаев Д.Н. Проблема психосоматических взаимоотношений // Методология и социология педиатрии. СПб., 1991.

Кант И. Критика чистого разума. Соч. Т. 3. М., 1964.

——— Пролегомены. Соч. Т. 4. Ч. 1. М., 1965.

Карнап Р. Значение и необходимость. М., 1959.

——— Философские основания физики. М., 1971.

Касавин И.Т., Сокулер Э.А. Рациональность в познании и практике. М., 1989. Рациональность как предмет философского исследования. М., 1995.

Кессиди Ф.Х. От мифа к логосу. М., 1972.

Ким В.В., Миляева С.Ю. Наука и вненаучные формы рациональности (к проблеме генезиса рациональности) // Рациональность науки и практики: Закономерности сближения. Свердловск, 1989.

Козлова М.С. Логика и реальность // Вопр. философии. 1965. № 9.

——— Проблема оснований науки // Природа научного познания. Минск, 1979.

——— Философия и язык. М., 1972.

Колдуэлл, Д. Оккультный мир Е.П. Блаватской. Сборник / Пер. с англ. М., 1996.

Коноплева Н.П., Соколик Г.А. Симметрии и типы физических теорий // Вопр. философии, 1972. № 1.

Конт О. Курс положительной философии. Т. 1. СПб., 1903.

Копылов Г.Г. Эзотерика, наука, экстрасенсорика: как можно обойтись без «лженаук» // Филос. науки, 2001, № 2.

Корольков А.А., Петленко В.П. Диалектика и теоретические проблемы медицины // Материалистическая диалектика в пяти томах. Диалектика природы и естествознания. Т. 3. М., 1983.

Коршунова Л.С., Пружинин Б.И. Воображение и рациональность. М. 1989.

Космические легенды Востока // авт.-сост. С.В. Стульгинскис. М., 1992.

Кравец А.С. Социокультурная ниша науки // Физика в системе культуры. М., 1996.

Крайнева Т.В. Теософия как философия // XXI век: будущее России в философс ком измерении: Материалы Второго Российского философского конгресса

(7–11 июня 1999 г.) Т. 4. Философия духовности, образования, религии. Ч. 2. Екатеринбург, 1999.

Кузнецов Б.Г. Спиноза и Эйнштейн // Эйнштейновский сборник. М., 1968.

Кузнецов В.Г. Герменевтика и гуманитарное познание. М., 1991.

Кулаков Ю.И. Синтез науки и религии // Вопр. философии, 2000, № 2.

——— Структура и единая физическая картина мира. // Вопр. философии. 1975. № 2.

Лакатос И. Доказательства и опровержения. М., 1967.

——— Фальсификация и методология научно-исследовательских программ. М.. 1995.

——— История науки и ее рациональные реконструкции // Кун Л. Структура научных революций. М., 2003.

Леви-Брюль Л. Первобытное мышление. М., 1930.

Лейбниц Г.В. Новые опыты о человеческом разуме. М.-Л., 1963.

Лобковиц Н. От субстанции к рефлексии. Путь западноевропейской метафизики // Вопр. философии. 1995. № 1.

Лосев А.Ф. Дерзание духа. М., 1989.

——— Философия, мифология, культура. М., 1991.

Макдауэлл Дж. Не просто плотник. М., 1992.

Мамардашвили М.К. Как я понимаю философию. М., 1992.

——— Сознание как философская проблема// Вопр. философии. 1990. № 10.

Мамчур Е.А. Анализ различных направлений см. в книге: Критика современных немарксистских концепций философии науки. М., 1987.

——— Проблема выбора теории. М.,1975.

——— Физика и этикет // Физика в системе культуры. М., 1996.

Марков Б.В. Наука как познание и форма жизни // Логика и развитие научного знания. СПб., 1992.

Маркова Л.А. О возможности соотношения науки и религии // Вопр. философии, 1997, № 11.

Маркс К. Капитал. Т. 1. // Маркс К., Энгельс Ф. Соч. Т. 23.

——— Тезисы о Фейербахе // Маркс К., Энгельс Ф. Соч. Т. 3.

Маслиева О.В. Истоки лирической поэзии (художественное сознание в Древней Греции) // Контексты культуры: антропологический и аксиологический аспекты. СПб., 2008.

Мах Э. Анализ ощущений и отношение физического к психическому. М., 1907.

Милль Дж. Система логики силлогистической и индуктивной. М., 1914.

Мудрагей Н.С. Рациональное и иррациональное. Историко-теоретический очерк. М., 1985.

Назиров А.Э., Маслиева О.В. Генезис духовной культуры и психосоматика // Медицина. Культура. Философия. СПб., 1994.

Назиров А.Э. Эволюционные взгляды в теософии // Взаимодействие науки, философии и богословия в формировании идей эволюции в природе и обществе. СПб., 1999.

Нарский И.С. Западноевропейская философия XVII века. М., 1974.

Ньютон И. Математические начала натуральной философии. Пг., 1916.

Панов В.Г. Эмоции, мифы, разум. М., 1992.

Парнов Е.И. Трон Люцифера: Критические очерки магии и оккультизма. М., 1991.

Петленко В.П., Погорелов В.И., Попов А.С. Медицина и эстетика. Эстетические нормы в медицине // Исторический материализм и медицина. Л., 1981.

Петров М.К. Язык, знак, культура. М., 1991.

Платон. Апология Сократа // Соч. в 3-х т. Т. 1. М., 1963.

Платонов К.И. Слово как физиологический и лечебный фактор. М., 1957.

Поппер К. Логика и рост научного знания. М., 2004.

Пуанкаре А. Наука и метод. СПб., 1910.

Раджабов У.А. От мифов к современным космологическим концепциям // Филос. науки. 1991. № 7.

Разумовский О.С. Закономерности оптимизации в науке и практике. Новосибирск, 1990.

Рассел Б. Человеческое познание. М., 1957.

Ребещенкова И.Г. Интерпретация естественных наук в герменевтике (критический анализ концепции П. Хилэна) // Филос.науки. 1987. № 4.

Ренан Э. Жизнь Иисуса. М., 1991.

Рожанский И.Д. Развитие естествознания в эпоху античности. М., 1979.

Розин В.М. Эзотерический мир // Общественные науки и современность, 1992, № 4.

Сандулов Ю.А. Христианство и культ дьявола // История религии. СПб., 1997.

Сергеев К.А., Силин Я.А. Диалектика категориальных форм познания (космос Аристотеля и наука Нового времени). Л., 1987.

Сизов С.С. О природе страха // Методология и социология педиатрии. СПб., 1991.

Сокращенный латинский словарь Ананьева, Яснецкого, Лебединского, изданный П.М.Леонтьевым. М., 1883.

Сорина Г.В. Психологизм и антипсихологизм о роли логики в научном познании (конец XIX—начало XX века) // Филос. науки. 1986. № 6.

Сочинения Декарта. Рассуждение о методе // Соч. в 2-х т. М., 1989.

Спиноза Б. Этика. Часть первая. // Избр. произв. В 2-х т. Т.1. М., 1957.

Столяр А.Д. Тезис К.Маркса о материальном “производстве идей” в свете археологических источников // Советская археология. М., 1983. № 4.

Тайлор Э. Первобытная культура. М., 1989.

Тахо-Годи А.А. Греческая мифология. М., 1989.

Токарев С.А. Ранние формы религии. М., 1992.

Тухолка С.В. Оккультизм и магия. Сердце № 1, СПб., 1992.

Фейерабенд П.К. Избранные труды по методологии науки. М., 1986.

Фейнман Р. Характер физических законов. М., 1988.

Филатов В.П. Живой космос: человек между силами земли и неба // Вопр. философии 1994. № 2.

Философский энциклопедический словарь // Гл. редакция: Л. Ф. Ильичев, П. Н. Федосеев, С. М. Ковалев, В. Г. Панов—М.: Сов. Энциклопедия, 1983.

Флоренский П.А. Вступительное слово пред защитою на степень магистра книги: "О духовной истине". Москва, 1912 г., сказанное 19-го мая 1914 г. // Флоренский П.А. Столп и утверждение истины (II). М., 1991.

Фрагменты ранних греческих философов. Сост. Лебедев А.В. Ч.1. М., 1989.

Франк Ф. Философия науки. М., 1960.

Фрейд З. Психология бессознательного. М., 1989.

Фромм Э. Душа человека. М., 1992.

Фрэзер Дж. Дж. Золотая ветвь. Исследование магии и религии. М., 1980.

Хаббард Л. Рон. Дианетика: Современная наука душевного здоровья. М., 1993.

Хайдеггер М. Работы и размышления разных лет. М., 1993.

Хейзинга Й. Homo ludens. Опыт исследования игрового элемента в культуре // Самосознание европейской культуры XX века. М., 1991.

——— Homo ludens. М., 1992.

Царегородцев Г.И., Ерохин В.Г. Диалектический материализм и теоретические основы медицины. М., 1986.

Целищев В.В. Логическая истина и эмпиризм. Новосибирск, 1974.

Шлик М. О фундаменте познания // Аналитическая философия. Избранные тексты. М., 1993.

Шпренгер Я., Инститорис Г. Молот ведьм. М., 1990.

Шри Ауробиндо Гхош. Основы йоги. СПб., 1993.

Шри Шримад А.Ч. Бхактиведанта Свами Прабхупада. Учение Шри Чаитанйи. Вильнюс-Москва-Ленинград-Калькутта-Бомбей-Нью-Дели, 1989.

Шталь И.В., Попова Т.В. О некоторых приемах построения художественного образа в поэмах Гомера и византийском эпосе XII в. // Античность и Византия. М., 1975.

Штофф В.А. Моделирование и философия. Л., 1966.

Шюре Э. Великие посвященные: Очерк эзотеризма религий. М., Репринтное воспроизведение издания 1914 г.

Элиаде М. Космос и история. М., 1987.

Юнг К.Г. Феномен духа в искусстве и науке. М., 1992.

Ясперс К. Истоки истории и ее цель // Ясперс К. Смысл и назначение истории. М., 1991.

Name Index

A

Apel 51
Aristotle 48, 53, 55, 67, 58
Aurobindo 7, 29

B

Bachelard 27
Bacon 55–56, 68
Barlet 38, 49
Bauer 3, 28
Blavatskaya 35, 39, 49, 40, 50, 41, 46–47
Bransky 34
Bridgman 69
Brinton 12, 31

C

Carnap 69, 62, 64–65
Chaitanya 33
Comte 12, 60, 69

D

Derrida 35
Descartes 20, 54, 56, 68, 59

E

Eliade 28, 33

F

Feyerabend 2, 66–67, 70
Feynman 68
Florensky 24, 34
Frank 62, 69, 64
Frazer 31

Freud 2, 28, 7, 29, 8, 55
Fromm 8, 29, 9, 55

G

Gadamer 27

H

Hegel 29, 32, 39, 50, 40, 51, 54, 57, 59, 69
Heidegger 27, 25, 35, 26, 55
Hobbes 56, 68
Hubbard 29
Huizinga 20, 33
Husserl 51, 55

I

Institoris 35

J

Jaspers 53, 67
Jung 7, 28–29, 55

K

Kant 20, 51, 58, 68, 59, 69
Kuhn 2, 34, 66, 70

L

Lakatos 34, 66–67, 70
Leibniz 54, 58, 68
Lévi-Bruhl 31
Locke 20
Losev 31, 33

M

Mach 60, 69, 63
Marx 28, 30, 55
Maslieva 28, 32
McDowell 28
Mill 60, 69
Mostepanenko 34

N

Narsky 68
Nazirov 28, 50
Newton 29, 54, 68

P

Petrov 31
Plato 19, 32, 20, 25, 38, 41, 48–49, 54–55
Poincaré 62, 69
Popper 65, 69–70, 67, 71

R

Reichenbach 61, 69
Renan 23, 34
Russell 61, 69

S

Schlick 64, 69
Schuré 32
Shtoff 69
Spencer 30
Spinoza 57, 68, 59

Sprenger. *See* Institoris 35
Srimad 33
Stepin 34

T

Tokarev 30–31
Tylor 30–31, 12, 21, 34, 22, 47

V

Vygotsky 34

W

Wittgenstein 61, 64, 69

Subject Index

A

a priori 57, 59
absolute 17, 47, 55, 58
abstract 6, 12, 22–23, 40, 47, 49, 57, 59
aesthetic 10–11, 19
Agni 31, 15–16, 20
Ahriman 13
animism 11–12, 21, 23, 45–46, 51, 47, 49
animistic 12, 21, 24–26, 38, 41, 43, 45–47, 51, 48–49, 53
animistic thinking vii, 38, 46–47
animizing 44, 48
apeiron 19, 53
arche 19
axiological 5, 23–24, 48
axiology 24

B

biosocial 3, 5

C

classical 22, 26, 39, 45, 51, 54, 56, 71
cognitive 1, 40, 61–62
concept 1–3, 28, 7, 9, 12, 19, 21, 23–24, 38, 43–44, 54–55, 57, 59–60, 62, 71
conceptual 7, 38, 48–49, 55
concrete 6, 55, 58–59
construct 22–23, 51, 59–60
construction 12, 19, 21, 44, 57, 60
continuity 67
continuous 45
cosmic 15, 49, 42–43
cosmos 16, 20, 33
cultural 1–2, 5, 7–10, 12, 19, 25–26, 51, 53
culture 1–2, 5–12, 31, 15, 17–20, 33, 21–22, 34, 23–26, 39, 46, 51, 47, 53–54

D

deduction 54, 57
deductive 55–56, 62

dialectic 8, 19
dialectical 54, 60
dialectics 55, 58
differential 54
displacement 19–20, 22, 25

E

eidos 19
empirical 55–58, 60–63, 65–66, 71
empiricism 7, 56, 59–61, 64
epic 13, 16–18
epistemological 20, 56, 61
epos 11, 16, 18–19, 22
esoteric 11, 34, 37, 40, 43, 45, 48
esotericism 49, 38, 50
extreme 16, 60

F

falsifiability 39, 66
fetish 31, 12, 20
First Cause, the 41, 43
Fohat 41–42
foundation 10
function 3, 29, 15, 34, 67
functional 2–3, 6, 9–10

G

generalization 24, 56
gnoseological 5–6, 11, 19, 21–23, 39, 46, 48, 65, 67, 71
gnoseology 24, 63–64
grammar 25

H

hermeneutic 2
hermeneutics 2, 27
historical 1–2, 5, 9, 18, 25, 39, 54
history 6–8, 16, 33, 22, 24, 45, 55

I

idea 4, 30, 19, 33, 23, 27, 39, 41–44, 48, 54, 56, 60
ideal 3, 6–7, 9, 11–12, 19, 21, 24–26, 38, 41, 45, 51, 48–49, 54–55, 71
impersonal 11, 41
individual 2, 6, 8–9, 11, 13, 34, 40, 44–45, 58, 64
inductive 56, 60
integral 5, 11, 54
intuition 58, 62
irrational 1, 22, 34

L

logic 25, 42, 53–55, 58, 60–62
logical 5, 30, 24–25, 38, 49, 53–54, 57–58, 60–66, 71
Logos, the 41, 44

M

material 3, 5, 9–10, 31, 13, 34, 26, 38, 40–41, 43–44, 46–47, 55, 71
mathematical 54–55, 60–61, 64, 71
mathematics 27, 53–55, 60, 62, 65
Matter 41–46
meaning 6, 22, 38–39, 45, 58, 61–62
metaphysical 22, 26, 53, 59, 61, 65
metaphysics 19, 54, 60
methodological 1–2, 4, 26–27, 39, 58, 63, 67
methodology 1–2, 4, 55, 68, 57–58, 60, 62, 64, 71
mind 6, 29, 14, 20, 24, 40, 45, 47, 58, 62
minimum 54
model 3, 16, 19, 22–24, 55–56, 62, 64, 71

modern European science 25, 48, 54
moral 15, 33, 23–24, 41
morality 18, 23, 56
mystical 8, 30, 20, 37, 48
mysticism 44
myth 12, 20

N

nominalism 54
non-classical 22, 25–27, 51, 54, 71
norm 3, 28, 4–5, 7–9, 20
normal 7–8, 67

O

object 2, 6–7, 9, 11–13, 22, 26–27, 40–41, 54, 56, 62, 67, 71
objective 26, 54, 71
occult 5, 37–38, 49, 39
occultism 37–38, 49, 44

P

Parabrahman 43
paradigm 27, 67
pathological 3–4, 8
pathology 3–5, 8
patristics 55
phenomenon 4, 6, 9, 12, 26, 51, 49
philosophical 8, 18–20, 22, 25, 27, 38, 50, 41–42, 47, 51, 48–49, 53, 56, 58–61, 65–66, 71
philosophical thinking 47, 51, 48–49
philosophy 2, 11, 19–22, 24, 35, 26, 37–41, 45–47, 51, 48–49, 53–55, 57–61, 66, 71
poetic 19–22, 25, 38, 48, 53
poetization 25
polytheism 13, 17
position 1, 4, 34, 22, 26, 42, 54, 61, 63, 65
positivism 60–66
positivist 60
post-positivist 2, 67
Prakriti 41
principle 3, 7, 30, 17, 19, 27, 42, 44, 54–55, 58–61, 63, 65–66
program 39, 60, 65, 67
psychogenic 5, 13
psychosomatic 2, 4–5, 7–9, 29, 10, 30, 20
psychosomatics 1–2, 5, 7, 9, 20
Purusha 20, 41

R

rational 1–2, 30, 13, 17, 22, 24–25, 37, 41, 56, 58, 70
real 5, 29, 21, 23, 54, 58
realism 54
reason 48, 54, 58–59
reflect vii, 50, 63
reflecting 17, 20, 47
reflection 11, 19–20, 25, 40, 45–46, 51, 48, 53, 55, 59, 64
reflective 25, 42
reflexive vii, 18, 40
rhetoric 25
rhetorical 48

S

scheme 54, 65
scholastic 39
scholasticism 54–55
scientific 1–2, 28, 22, 25, 42, 53–55, 68, 56–62, 64–67, 71
scientific thinking 25, 53
sign 11–12, 31, 38, 45, 64
soma 16
speculation 41, 62
speculative 22–24, 42, 44, 53, 60
spirit 13–14, 31, 20, 34, 40–41, 44–45, 54
spiritual 1–3, 5–7, 9–13, 15–18, 33, 21–23, 34, 24–26, 38–39, 41, 43–44, 46–47, 51, 53

subject 2, 27–28, 26, 41, 48, 54–56
subjective 26, 37, 48, 54, 71
substance 15, 19, 39, 41, 43, 46, 57, 67
synthesis 2, 35, 26, 37, 39, 43, 47, 56
synthetic 59

T

theoretical 2, 4, 37, 53, 56–58, 60–61, 65–67
theory 2–3, 7, 12, 27, 49, 54–67, 71
theosophical 43–48
theosophy 37, 39, 50, 45–49
thinking 9, 30, 19, 22, 34, 25–27, 38–40, 45–47, 51, 48–49, 53–57, 59–60
transcendental 58

U

understanding 2, 7, 29, 10, 30, 12, 19, 34, 37–38, 40, 47, 59, 63
unmanifested 41

V

verification 39, 61
verify 63–64

W

worldview 24, 40, 47

PETER LANG PROMPT

Peter Lang Prompt is a new product that offers our authors the possibility of publishing their research in its intended form, outside the traditional length of a standard book or a journal article. It is produced within a faster production time, without missing the high standard of scholarly publishing. This concise model gives researchers the chance to publish their time-sensitive research, open a forum for debate, and create a greater impact.

For further information, please contact:

Peter Lang Publishing, Inc.
Acquisitions Department
29 Broadway, 18th floor
New York, New York 10006

To order, please contact our Customer Service Department:

peterlang@presswarehouse.com (within the U.S.)
orders@peterlang.com (outside the U.S.)

Visit our website: www.peterlang.com

Prompts include:

Claudia Aburto Guzmán, *Poesía reciente de voces en diálogo con la ascendencia hispano-hablante en los Estados Unidos: Antología breve* (978-1-4331-5207-8)

Janet Farrell Leontiou, *The Doctor Still Knows Best: How Medical Culture Is Still Marked by Paternalism* (978-1-4331-7322-6)

Josiane Ranguin, *Mediating the Windrush Children: Caryl Phillips and Horace Ové* (978-1-4331-7424-7)

Dylan Scudder, *Coffee and Conflict in Colombia: Part of the Pentalemma Series on Managing Global Dilemmas* (978-1-4331-7568-8)

Dylan Scudder, *Conflict Minerals in the Democratic Republic of Congo: Part of the Pentalemma Series on Managing Global Dilemmas* (978-1-4331-7561-9)

Dylan Scudder, *Mining Conflict in the Philippines: Part of the Pentalemma Series on Managing Global Dilemmas* (978-1-4331-7632-6)

Dylan Scudder, *Multi-Hazard Disaster in Japan: Part of the Pentalemma Series on Managing Global Dilemmas* (978-1-4331-7530-5)

Angela Williams, *Hip Hop Harem: Women, Rap and Representation in the Middle East* (978-1-4331-7295-3)

Ivan Zhavoronkov, trans., *The Socio-Cultural and Philosophical Origins of Science* by Anatoly Nazirov (978-1-4331-7228-1)